Roshan G. Ragel
Sri Parameswaran

Microarchitectural Support for Security and Reliability

Roshan G. Ragel
Sri Parameswaran

Microarchitectural Support for Security and Reliability

An Embedded Systems Perspective

VDM Verlag Dr. Müller

Imprint

Bibliographic information by the German National Library: The German National Library lists this publication at the German National Bibliography; detailed bibliographic information is available on the Internet at http://dnb.d-nb.de.

Cover image: www.purestockx.com

Publisher:
VDM Verlag Dr. Müller Aktiengesellschaft & Co. KG , Dudweiler Landstr. 125 a, 66123 Saarbrücken, Germany,
Phone +49 681 9100-698, Fax +49 681 9100-988,
Email: info@vdm-verlag.de

Zugl.: Sydney, University of New South Wales, 2006

Produced in USA and UK by:
Lightning Source Inc., La Vergne, Tennessee, USA
Lightning Source UK Ltd., Milton Keynes, UK
BookSurge LLC, 5341 Dorchester Road, Suite 16, North Charleston, SC 29418, USA

ISBN: 978-3-639-01472-3

To the memories of all my Teachers...

PREFACE

Security and reliability in microprocessor based systems are concerns requiring adroit solutions. Security is often compromised by code injection attacks, jeopardizing even trusted software. Reliability is of concern, where unintended code is executed in modern processors with ever smaller feature sizes and low voltage swings causing bit flips. Countermeasures by software-only approaches increase code size and therefore significantly reduce performance. Hardware assisted approaches use additional hardware monitors and thus incur considerably high hardware cost and have scalability problems. Considering security and reliability issues during the design of an embedded system has its advantages as this overcomes the limitations of existing solutions.

This book is written for hardware design engineers and students with fundamental knowledge of VLSI logic design. The benefits of considering security and reliability issues during the design of an embedded system can be exploited by designers with expertise in the field of VLSI design, computer architecture and embedded systems. This book provides essential knowledge on these areas and focuses on the practical solutions for real-world problems.

This book constitutes of the PhD thesis of the first author, which has been performed at the School of Computer Science and Engineering at University of New South Wales, Sydney during the period between 2003 and 2006. The material presented in this book combines two elements: one, defining a hardware software design framework for security and reliability monitoring at the granularity of micro-instructions, and two, applying this framework for real world problems.

In Chapter 1, the authors introduce the common security and reliability problems and have supply evidences for the severity of these problems; and therefore the need for solutions to such problems in embedded systems environment. In Chapter 2, they provide a detailed analysis of the known security and reliability issues in general and a complete study on different types of code injection attack. In Chapter 3, they review the literature for the proposed countermeasures for the security and reliability issues reported in Chapter 2.

In Chapter 4, the authors introduce the security and reliability framework, which was developed during the first author's research programme in compliance with his PhD. In Chapter 5, they explain and show the overheads of an inline security and reliability monitoring technique that counters common security and reliability problems. In Chapter 6, they detail a technique to detect code integrity violation, known as encrypted basic block check-summing, to counter the most widely encountered software based security threat, known as code injection attacks. The authors have also discussed the applicability of the same technique to reliability assurance, and report the overheads and error coverage of this approach.

In Chapter 7, they propose a hardware assisted control flow checking method to preemptively detect control flow errors in applications due to soft errors. Further, the overheads of this approach and a comparison of this method against a software only approach are discussed. In Chapter 8, they propose an extension to the control flow checking technique proposed in Chapter 7. The extension will cover soft errors in the program counter and the register file.

Acknowledgements

I have been fortunate enough to meet and be with many wonderful people over the years, who have given me their time, companionship, friendship and professional, moral and personal helps. It is time to recall and acknowledge all of them at this special moment of time.

I would first like to thank Professor Sri Parameswaran, the co-author of this book, for supervising my PhD Thesis. He not only gave me the technical support and supervision that a graduate student could expect from his supervisor, but he also encouraged and gave moral support without which I would never have made it this far.

A list that, alas, has far too many names on it to mention separately is that of all researchers and colleagues at Embedded Systems Laboratory (ESL), University of New South Wales - my former working place. It was pleasure, fun and stimulating of being in such an environment and I am definitely sure that missing them all. All the insightful conversations we exchanged over the past couple of years had made my stay at ESL a delightful experience.

Thanks go out also to my thesis examiners, for their time that they spent on reviewing my writing, their valuable feedbacks and comments and their efforts on evaluating it.

I would like to thank my parents and my brother, Chester, for their unconditional love, patience and support. I definitely missed them all during my stay in Sydney to pursue my PhD. I also would like to thank all my friends especially Dr. S. Radhakrishnan for their help, friendship, understanding and patience.

Finally, I would like to extend my thanks to all the researchers around the world, who are working on computer architecture, security and reliability whose work have motivated me in many aspects. Particularly, all the reviewers and editors of my papers for their priceless efforts and feedbacks.

Roshan G. Ragel
April, 2008

Table of Contents

Chapter 1

Introduction

... 'begin at the beginning and go on till you come to the end: then stop!'...

— Lewis Carroll, *Alice's Adventures in Wonderland*

Considering reliability and security issues during the design of an embedded system has its advantages as this overcomes the limitations (such as the performance and code size overheads in the software-only techniques and scalability problems in the hardware assisted techniques) of existing solutions.

1.1 About this Chapter

1.1.1 Objectives

This chapter has three primary goals and they are:

1. providing evidence for increasing security and reliability problems in computing systems;

2. showing the need for security and reliability research in embedded systems.

1.1.2 Contributions

Apart from introducing and summarizing this book, this chapter provides statistics to motivate the work, that shows the impact of security and reliability problems.

1.2   Embedded Systems and Security

Security is becoming an important consideration in the design of general purpose and embedded processors. As we move forward in the digital communication age, not only are the high end systems such as network routers, gateways, firewalls, and storage, web and mail servers connected to the Internet, but so are many low end electronic devices such as personal computers, smart mobile phones, personal digital assistants, smart cards, etc. which need to store, compute and forward sensitive information. Technology advancement which has pushed the development of these devices, and the explosive growth of the Internet have both increased and refined the malicious attacks that these systems encounter. It has been noticed that the cost we have to pay for ignoring security in systems is very high. For instance, a recent survey [106] on computer crime and security by Computer Security Institute (CSI) and Federal Bureau of Investigation (FBI) released in 2006*, reports that just the organizations connected to the individual participants in the survey had lost more than half a billion dollars due to security attacks. Therefore, successful attacks could bring widespread and devastating consequences due to our increasing dependence on networked information systems. Therefore, security threat has become one of the major constrains that prevents users from adopting next generation data applications and services. This claim could be supported by another recent survey conducted by ePaynews, which reveals that around 50% of mobile phone and PDA [12, 228] users are concerned about security threats which prevent them from using these devices for mobile commerce.

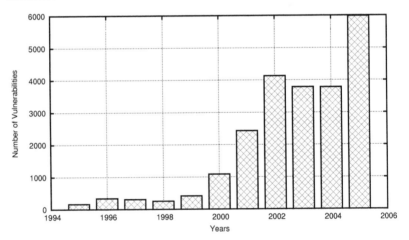

Figure 1.1: *Number of unique security vulnerabilities reported over the past years (source [44])*

Integrity, availability and secrecy are crucial for military applications, and ever-growing numbers of businesses and individuals who use networked devices. With the evolution

---

* The information is gathered for the calender year 2005 from 616 computer security practitioners in US corporations, government agencies, financial institutions, medical institutions and universities

of the Internet, an additional wide variety of security issues, such as authentication, privacy, denial of service, digital content protection, etc. need to be looked at. Security of processors have been the subject of extensive research in connection to computing and communications systems. Outcome of this research includes theoretical improvements in cryptography and security protocols, and standards such as WEP[†] (Wired Equivalent Privacy), IPSec[‡] (Internet Protocol Security), SSL[§] (Secure Sockets Layer), WTLS[¶] (Wireless Transport Layer Security), etc. [34, 317]. While theoretical improvements are essential in the area of security, secure implementations have to play a vital role since security attacks also take advantage of weaknesses in system implementations [11]. Therefore, a part of this thesis addresses secure implementations of embedded processors to counter common security attacks.

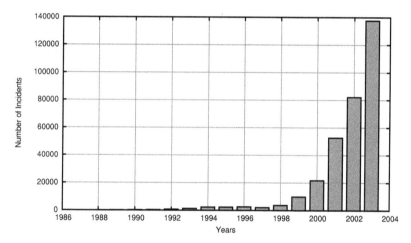

Figure 1.2: *Number of security incidents reported over the past years (source [44])*

The elementary solution to protect against malicious code execution on embedded processors is to write software that is not vulnerable to such attacks. Although a thorough static analysis may reveal a significant number of these vulnerabilities, it is practically impossible to write code that is not vulnerable. This is clearly apparent by the ever increasing number of vulnerabilities reported over the years. For example, Figure 1.1 depicts the number of unique security vulnerabilities reported for every calendar year from 1994 to 2005. This information is taken from CERT/CC[||] (CERT Coordination Center) Statistics

---

[†] the privacy protocol specified in IEEE 802.11 to provide wireless LAN users protection against casual eavesdropping

[‡] a standard for securing Internet Protocol communications by encrypting and/or authenticating all IP packets. It provides security at the network layer.

[§] a universal security protocol used mainly to secure communication with web servers and therefore with e-business and Internet banking applications

[¶] a part of Wireless Application Protocol - WAP aims to provide Internet content and advanced telephony services to digital mobile phones, pagers and other wireless terminals

[||] CERT is a center of Internet security expertise, located at the Software Engineering Institute, a federally funded research and development center operated by Carnegie Mellon University, in the US

1988-2006 [44]. As it is shown by the bar chart in Figure 1.1, the number of vulner-abilities reported over the years is increasing drastically. To be included on the report, vulnerabilities must meet the following five requirements and they are (1) they affect a large number of users, (2) they have not been patched on a substantial number of systems, (3) they allow computers to be taken over by a remote, unauthorized user, (4) sufficient details about the vulnerabilities have been posted to the Internet to enable attackers to exploit them, and (5) they were discovered or patched for the first time, after the attacks based on these vulnerabilities become common.

Figure 1.2 depicts the number of security incidents reported over the years from 1988 to 2003 as reported by [44]. The number of incidents have been increasing exponentially over the years. CERT/CC has stopped publishing the number of incidents as of 2004 and the reason for this act quoted as:

> *'Given the widespread use of automated attack tools, attacks against Internet-connected systems have become so commonplace that counts of the number of incidents reported provide little information with regard to assessing the scope and impact of attacks. Therefore, as of 2004, we will no longer publish the number of incidents reported.'*

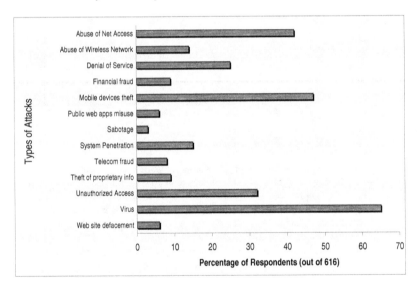

Figure 1.3: *Security Attacks in calendar year 2005 by Types (source [106])*

Figure 1.3 illustrates the different types of security attacks encountered by a set of security practitioners in the US in 2005. This information is obtained from *the CSI/FBI Computer Crime and Security Survey, 2006* [106]. The survey is based on information collected from 616 individuals and is based on their experience in the calender year 2005. The top five types of attacks from the list are: virus attacks, theft of mobile devices, abuses of Net access, unauthorized access of systems and denial of service attacks.

Figure 1.4 shows the financial losses in a set of organizations in the US due to different types of security attacks in 2005 [106]. Virus attacks are found to be the source of the greatest financial losses. Unauthorized access ranked as the second-greatest source of financial loss. Financial losses related to laptops (or mobile hardware) and theft of propri- etary information (i.e., intellectual property) are third and fourth. These four categories account for around 75% of the total financial losses.

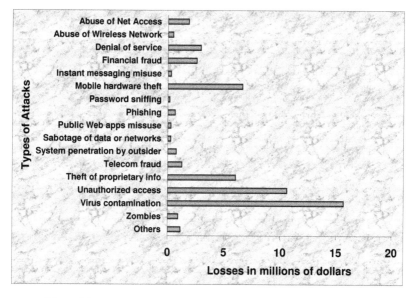

Figure 1.4: *Losses in millions of dollars due to Security Attacks (source [106])*

The use of embedded systems everywhere to capture, store, manipulate and access sen- sitive data and information has given the system several strange and interesting security demands. The increasing number of occurrences of security breaches over the years is the main reason for the work towards secure embedded processors. Embedded system designers have often mis-understood security in the context of embedded systems as the addition of features to the system such as specific cryptographic algorithms and security protocols [146]. Kocher et al. in [146] argue that the security in embedded processors has to be taken as an entirely new metric that should be considered throughtout the design process, along with the other metrics such as performance, area and power. We agree with the authors of [146] and therefore, in this book present how design time solutions could be implemented in embedded processors to provide security.

Software based security attacks could generally be divided into violating software in- tegrity and violating program data properties. Most of the recent security attacks result in demolishing code integrity of an application program [184, 185]. This includes dy- namically changing instructions with the intention of gaining access to a program. The design time solutions implemented during this book for secure embedded processors are proposed to counter software based security attacks, that violate software integrity of applications.

1.3 Embedded Systems and Reliability

Transient faults (also known as soft errors) corrupting the memory content were reported as early as 1954 [125]. In the late seventies, a large number of landmark papers revealed that single even upsets[**] (SEU) were introduced in semiconductor memories by cosmic rays [112, 330] and radioactive contamination [175]. However, the recent deployment of embedded systems in complex mission and life-critical environment has increased the significance and impact of these errors. Besides, the need for deployment of concurrent error detection (along with security) in embedded processors is increasing as the device geometry decreases. Devices become more susceptible to transient faults as clock frequency increases and difference between voltage level decreases. The voltage levels of current microprocessors are so low such that cosmic rays could alter voltage levels representing data values [231]. This continuously decreasing device sizes and supply voltages has made even the flip-flop circuits (which were considered unsusceptible to soft errors in early days) susceptible to transient errors and has increased the soft error rates (SER) experienced by computing systems. Furthermore, increasing number of embedded processors are subjected to noisy environments (as the embedded systems are becoming more common in automobiles, public transport systems, mobile devices, process engineering industry, etc. [242]), which encourage transient faults.

Figure 1.5 depicts the estimated SER contributions of various elements for typical designs such as microprocessors, network processors, etc. The information in the figure is reported by researchers from Intel [189] in year 2005. It could be seen from Figure 1.5, that the combined SER contribution of sequential elements and combinational logic exceeds that of the unprotected SRAMs[††] (static random access memory). This finding goes against the long time claims of the researchers in the past, that the SEU is a phenomenon largely related to the memory modules. Therefore, special attention is required to develop techniques for protecting hardware components, which are not SRAM portions of a design, from soft errors.

Reliability data for critical systems are rarely published [161]. However, few available recent studies [24, 156, 259] reveal that the SER of a single bit, for both DRAM[‡‡] (dynamic random access memory) and SRAM are decreasing as shown in Figures 1.6 and 1.7

[**]A single event upset is a change of state, or voltage pulse caused when a high-energy particle strikes a sensitive node in a micro-electronic device, such as in a microprocessor, semiconductor memory, or power transistors. An error in device output or operation caused as a result of a single event upset is considered a soft error.[261]

[††]SRAM is a type of semiconductor memory. The word 'static' indicates that the memory retains its contents as long as power remains applied, unlike dynamic RAM that needs to be periodically refreshed [276].

[‡‡]DRAM is a type of random access memory that stores each bit of data in a separate capacitor. As real-world capacitors are not ideal and hence leak electrons, the information eventually fades unless the capacitor charge is refreshed periodically. Because of this refresh requirement, it is a dynamic memory as opposed to SRAM and other static memory. Its advantage over SRAM is its structural simplicity: only one

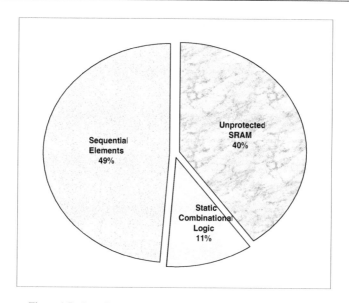

Figure 1.5: *Contributions to the overall SER of a design (source [189])*

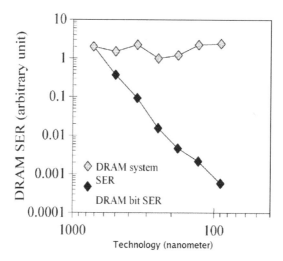

Figure 1.6: *SER in DRAM with the Technology scaling (taken from [24])*

respectively (Note that the units used to measure SER in the graphs are arbitrary, as the

---

transistor and a capacitor are required per bit, compared to six transistors in SRAM. This allows DRAM to reach very high density [77].

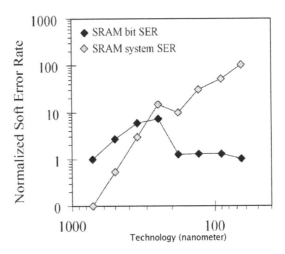

Figure 1.7: *SER in SRAM with the Technology scaling (taken from [24])*

authors of the work only wanted to show the comparision and not the values). However, the SER of DRAM and SRAM components are driven by Moore's law, and therefore the study shows an increase in SER with the advance in technology. Further, the authors of [259] have reported a flat single bit upset (SBU) rate for bits in combinational and sequential logic circuits. Therefore, the system SER for logic circuits will be increasing faster than the memory modules. Another observation from [259] is that the multi bit upset (MBU) is increasing steeply over the years with advanced technologies.

Even though the amount of SER information available in the literature is limited, from what is available, it is reasonable to conclude that the SER is increasing in memory and a new trend of soft errors are emerging in logic circuits. Therefore, in a part of this book, we design and implement solutions to detect both SBUs and MBUs in memory and selected logical components which are exposed to errors in embedded processors.

System designers have considered different levels within the system such as device/logic level, circuit level and architectural level for managing transient errors and have proposed various redundancy mechanisms (such as hardware, time, information or a combination of these) for deployment. Logic level designs usually use a redundant or a self-checking circuit to detect and recover from transient errors. Circuit level designs use techniques such as transistor resizing, implementing enough functional margins to anticipate for errors and forward body bias for detecting and recovering from transient errors. Examples for architecture level technique [265] are replication of application execution and duplication of functional units.

Reliability and Availability Measures

$$MTBF = \frac{\text{Total Elapsed Time} - \text{Sum of Downtime}}{\text{Number of Failures}} \tag{1.1}$$

$$Availability = \frac{\text{Total Elapsed Time} - \text{Sum of Downtime}}{\text{Total Elapsed Time}} \tag{1.2}$$

One of the often used reliability measures of a system or a component is, mean time between failures (MTBF), which is defined by Equation 1.1. MTBF measures the probability of failure for a component of a system, which is the general measure of a system's reliability. MTBF, usually measured in thousands or tens of thousands of hours, is the average time interval that elapses before a failure. Availability of a system is the percentage of time that a system is capable of serving its intended function and is measured using Equation 1.2.

Availability	Percentage	8-hour day	24-hour day
Two nines	99%	29.22 hours	87.66 hours
Three nines	99.9%	2.922 hours	8.766 hours
Four nines	99.99%	17.53 mins	52.60 mins
Five nines	99.999%	1.753 mins	5.260 mins
Six nines	99.9999%	10.52 seconds	31.56 seconds

Availability is usually measured in 'nines'. For instance, a system which has the availability level of 'five nines' is capable of supporting its intended function 99.999% of the total elapsed time. That is, in a system that works for all year 24x7, an availability level of 'five nines' represents an annual downtime of 5.26 [a] minutes. Table above tabulates the annual downtime in hours for the most commonly used availability percentages. The hours in the second column are for 8-hour days and in the third column are for 24-hour days.

---

[a]60x24x365.25x(1-99.999/100)

All the techniques used in different levels of a system, manipulate one or a combination of the redundancies mentioned in the previous paragraph. Hardware redundancy is accomplished by performing the same computation on multiple, separate hardware at the same time and validating the redundant results to disclose errors. When more than two redundant hardware is deployed, a majority voting scheme is deployed to find errors. Hardware redundancy schemes are good for computation with strict deadlines. However, these schemes suffer from common-mode failures[*]. Time redundancy is used when the cost of hardware redundancy is undesirable due to large hardware overheads. Time redundancy is usually obtained by running the same computation multiple times on the same hardware. Even though, time redundancy will not cost extra hardware apart from some hardware for collecting and comparing of multiple executions, it has high performance overhead. Information redundancy is usually used in memory and cache protections in

---

[*]same fault affecting multiple hardware components in the same way, resulting the errors unnoticeable

the form of parity bits [286] or error correction codes (ECC). Information redundancy requires additional hardware to store the redundant information and extra logic to manipulate them.

Fault injection studies conducted by Ohlsson et al. [206] and Schutte et al. [254] show that control flow errors due to transient faults comprise between 33% and 77% of all errors that occur in a computer system. Given that the majority of the system errors are transient (it is estimated that transient faults occur 10x to 30x more frequently than permanent faults, that is more than 90% of the failures in computers are due to them [267, 268]) and are not reproducible [163], a runtime error detection mechanism is the only feasible solution to detect control flow errors. The typical design parameters of a control flow checking solution are: error detection coverage; error detection latency; processor performance; memory overhead; and the monitoring hardware complexity [202].

Control flow errors are usually detected by dynamic software monitoring where additional software code is inserted into application programs; or, hardware-assisted runtime monitoring, where a hardware block is dedicated to perform security and reliability checks (for example a watchdog processor for control flow checking in [168]). Software only approaches increase code size enormously by adding additional instructions to perform monitoring and therefore significantly reduce performance. Furthermore, software only approaches only check for control flow errors within the user code and not library functions, unless the library functions are instrumented separately. Hardware assisted approaches use additional hardware blocks to perform monitoring and therefore incur very high hardware cost and need self checking mechanisms within the monitors to ensure that the monitoring hardware themselves are reliable.

Given the significance of control flow errors, a part of this book propose hardware-assisted control flow checking techniques in embedded processors to detect transient errors. We propose that considering control flow checking as a design parameter during the hardware design process of an embedded processor, will only slightly degrade the processor's performance, and cost little in hardware overheads.

1.4   Summary of this chapter

In this chapter, the authors have shown statistical evidence and discussed the need for secure and reliable embedded systems and therefore the need for the work presented in this book.

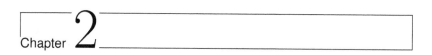

Chapter 2

Security and Reliability Issues

... 'and what is the use of a book,' thought Alice, 'without pictures or conversations?'

— Lewis Carroll, *Alice's Adventures in Wonderland*

M ajor security threats and reliability problems for embedded processors are discussed in this chapter. This chapter sets out the background context in which the thesis presented in this book is developed and understood and the motivation behind the projects involved in this book.

2.1 About this Chapter

2.1.1 Objectives

This chapter has the following primary goals.

1. This chapter details the major security threats that exist in embedded processors due to code injection attacks.

2. It introduces emerging reliability issues in embedded processors.

3. It sets out the background context in which the thesis is to be developed and understood.

2.1.2 Outline

Dependability of a computing system is defined as the trustworthiness of the system which allows reliance to be justifiably placed on the service it delivers. (The original definition of dependability is the ability to deliver service that can justifiably be trusted.) [16, 17]. Dependability includes the following attributes of a computing system:

1. Availability: readiness for correct service;

2. Reliability: continuity of correct service;

3. Safety: absence of catastrophic consequences on the user(s) and the environment;

4. Security: the concurrent existence of (a) availability for authorized users only, (b) confidentiality, and (c) integrity.

A system failure is defined by the behavior of the system that diverges from that prescribed by the user.

2.1.3   Contributions

Apart from touching most of the major security and reliability problems arising in Embedded Processors, this chapter details specific insights on the methods and techniques of code injection attack.

2.2   Security Issues

Security threats in embedded systems could be classified into privacy, integrity and availability attacks, based on the objectives of the attacks or into physical, side channel, and logical attacks based on the means used to launch the attacks [227, 228]. While privacy attacks could further be divided into authenticity, access control and confidentiality attacks, logical attacks could be classified into software and cryptographic attacks. Figure 2.1 explains how the attacks based on objectives are related to the attacks based upon means used to launch them. Figure 2.2 briefly explains the technical terms used in Figure 2.1.

Figure 2.1 further depicts examples for each type of security attacks. Examples for physical attacks include microprobing, reverse engineering and eavesdropping. As far as we are aware, all the physical attacks are privacy attacks as depicted by the dotted connection on Figure 2.1. The resources available for reverse engineering increase significantly if someone with manufacturing knowledge were to attempt to maliciously compromise the system. Integrated circuits may be vulnerable to microprobing or analysis under an electron microscope, once acid or chemical means have been used to expose the bare silicon circuitry [214]. Eavesdropping is the intercepting of conversations by unintended recipients which are performed when sensitive information is passed via electronic media, such as email or instant messaging.

As depicted in Figure 2.2, fault injection attacks [29, 30], power analysis attacks [171] (Simple Power Analysis - SPA and Differential Power Analysis - DPA, [145]), timing analysis attacks [35] and electro magnetic analysis attacks [224] are examples of side channel attacks. Side-channel attacks are performed based on observing properties of the system while it performs cryptographic operations. Countermeasures to side channel attacks, such as current flattening [190], randomized masking [278], and non-deterministic processing (scheduling) [123] are proposed in the literature.

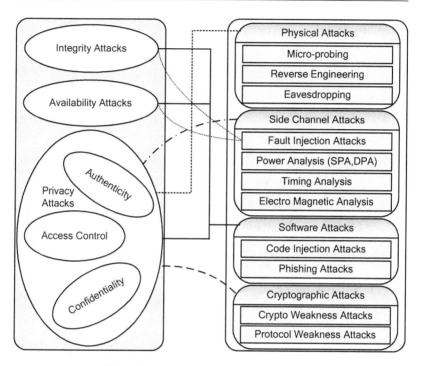

Figure 2.1: *Security Attacks on Embedded Systems*

Code injection attacks*, and phishing attacks are examples of software attacks. Majority of attacks which take place today comprise of code injection and phishing attacks. Code injection attack is futher discussed in Section 2.3. Phishing attacks attempt to bogusly receive sensitive information such passwords by imitating a trustworthy person in an electronic communication, such as email. Phishing attack is used to gather useful information such as internet banking details, etc. and is increasing in numbers in recent time according to a report from Anti-Phishing Working Group (APWG) [7] (as shown in Figure 2.3 taken from [6]). Interested readers are referred either to APWG's home page [7] or to the Wiki page on phishing [212] for further details and examples of phishing.

Cryptographic attacks (in Figure 2.1) exploit the weakness in the cryptographic and protocol information, to perform security attacks, such as breaking into a system by guessing the password. A short list of common crypto and protocol vulnerabilities is given in [228]. Solutions proposed in the literature to counter cryptographic attacks include run-time monitors that detect security policy violations [142] and the use of safety proof carrying code [197].

---

*Attacks violating code integrity are called code injection attacks, as they insert harmful instructions into the dynamic program stream

Integrity	breaking the confidentiality and then changing whole or part of data or code
Availability	making the whole or a part of the system unavailable to the normal functioning
Privacy	gaining sensitive information by spying an embedded system
Authenticity	creating and sending data to users by faking themselves as authentic and get back sensitive information
Access Control	gaining access to computing resources by faking as legitimate users
Confidentiality	active or passive unauthorized access of any data
Physical	attacks which require physical invasion into the system
Side Channel	attacks based on observing properties of a system while it is performing cryptographic operations
Software	attacks initiated via software applications
Cryptographic	attacks which target on weak cryptographic and protocol information

Figure 2.2: *Description of the types of Security Attacks*

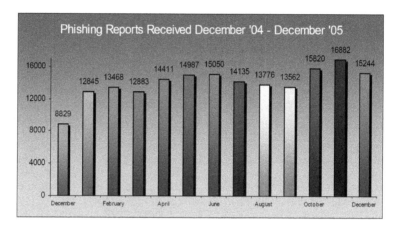

Figure 2.3: *Phishing Report (taken from [6])*

Inquisitive readers are referred to [146], [172] and [228] for interesting articles on embedded system security.

2.3 Code Injection Attacks

Most of the recent security attacks result in demolishing code integrity of an application program [185]. They include dynamically changing instructions with the intention of gaining control over a program execution flow. Attacks that are involved in violating software integrity are called code injection attacks. Code injection attacks often exploit common implementation mistakes in application programs and are often called security vulnerabilities. The number of malicious attacks always increase with the amount of software code [42, 43]. This subsection of the book details the most common security vulnerabilities present, which could lead to code injection attacks.

2.3.1 Stack-based buffer overflows

An array declaration in a programming language, such as C allocates a space in memory and the array is manipulated by a pointer to the first byte of the array. Almost all C compilers generate code that will not verify the size of the array when copying data into the array at runtime, and furthermore the runtime code will not carry the array size information. This allows programs to copy data further past the end of an array overwriting memory spaces not allocated to the array. When control flow information is stored in the adjacent memory slots to the array, it is possible for an attacker to overwrite it and gain access of the execution flow. Often this is the case with the stack, as it stores the address to return, after a function has been called. In most of the architectures the stack grows down the memory; that is, newer data in the stack would be pushed into a space addressed lower compared to an older data. The stack is divided into stack frames, one for each function and the top frame in the stack belonging to the current function. The introduction of stack frames brings the need for a frame pointer (which points to the starting address of the current stack frame) apart from the stack pointer (which points to the top of the stack). Each stack frame stores information about the corresponding function and they are, arguments to the function call, return address of the function, local variables, the last frame pointer and values of registers which are changed within the function.

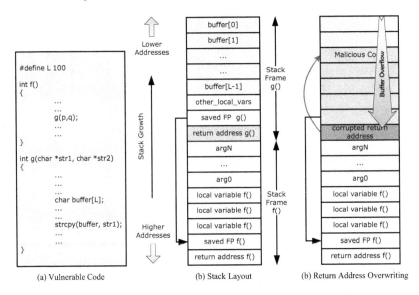

(a) Vulnerable Code   (b) Stack Layout   (b) Return Address Overwriting

Figure 2.4: *Stack based buffer overflow attack overwriting return address*

Figure 2.4 shows how a piece of vulnerable code could be exploited to perform return address overwriting. Figure 2.4(a) depicts a piece of C code which has function *f()* that calls another function, *g()* which is vulnerable. Function *g()* is vulnerable because it consists of an operation - *strcpy* which copies the string *str1* into *buffer* without considering the size of the string, *str1*. Since *buffer* is a local array, it will usually be stored in part of a stack frame belonging to local variables. A program that copies passing the end of this array, will overwrite anything stored after the array. Figure 2.4(b) depicts the program stack when function *g()* is executed. Function *g()* is called by function *f()* which has placed the arguments (*arg0..argN*) of function *g()* in the stack just after function *f()*s local variables before running a *call* instruction. The *call* instruction executed by function *f()* will write the return address (the address of the instruction next to the *call* instruction) into the stack. Now, function *g()* stores the current frame pointer into the frame pointer register and the old frame pointer into the stack. Function *g()* now has allocated space for its local variables in the stack where an array of characters for local variable *buffer* is allocated. Figure 2.4(c) depicts how the return address is overwritten using buffer overflow, by making use of a vulnerable function (*strcpy* in this case). Since *strcpy* does not check for the length of the string copied across, an attacker is able to make the program copy data beyond the end of an array (*buffer* in this case). Apart from the content of the *buffer*, the attacker has also overwritten the following pointers: the old frame pointer and the return address of function *g()*. In most of the attacks of this nature, the goal of the attacker is to overwrite the return address and this is the easiest route to gain control of an application programs execution flow. Usually, when a function returns, it will return to the address pointed by the return address in the stack. As described here, if the intruder has changed the return address to point to the code segment injected probably by the same copy operation (as in Figure 2.4(c)), when function *g()* returns, it will execute the injected malicious code and therefore giving the control to the attackers code [1, 196].

When overwriting the return address of a function by overflowing a buffer is restricted, either by a countermeasure or because an attacker could overwrite only a few bytes and cannot reach the return address in the stack, it is possible to exploit the stored frame pointer to gain control over the execution flow as depicted in Figure 2.5(c).

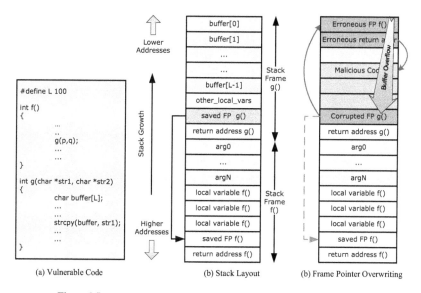

(a) Vulnerable Code   (b) Stack Layout   (b) Frame Pointer Overwriting

Figure 2.5: *Stack Based Buffer Overflow Attack Overwriting Frame Pointer*

As depicted in Figure 2.5, by overflowing the buffer[], an attacker could overwrite the saved frame pointer (*saved FP g()*) to point to a different address instead of the saved frame pointer of the previous stack frame. When function *g()* finishes execution, the old frame pointer value of function *f()* will be restored to the frame pointer register by popping the value from the stack of function *g()*. As in Figure 2.5(c), if the frame pointer in the stack belonging to function *g()* (*saved FP g()*) is corrupted by a buffer overflow, then the frame pointer for stack frame *f()* is going to be changed. This will ultimately result in returning of function *f()* to an attacker specified code segment as indicated by Figure 2.5(c) and therefore the intruder will gain control of the execution flow [143]. In case overwriting the return address and the frame pointer are protected by some counter-

measures, it is possible to use indirect pointer overwriting to gain access to the execution
flow of an application using buffer overflow attacks as described in Figure 2.6.

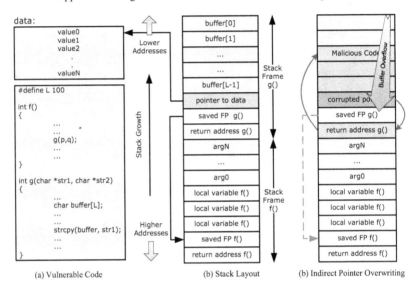

(a) Vulnerable Code   (b) Stack Layout   (b) Indirect Pointer Overwriting

Figure 2.6: *Stack based buffer overflow attack using indirect pointer overwriting*

As shown in Figure 2.6(a) and (b), a pointer to the data (a local variable to function *g()*
holding a pointer to *data0*) is stored in the stack frame belonging to function *g()*. By
overwriting this pointer to point to the return address of function *g()* and later changing
the value of the first word pointed by *data0* (i.e. *value0*), an intruder will be able to
return the execution flow of the application to a code segment pointed to by *value0* of
choice [36]. As depicted in Figure 2.6(c), overwriting pointer to *data0* to point to return
address *g()* and changing the value of *value0* to point to the malicious code injected by
the attacker the hijacking is achieved when the application de-reference return address
g(). Even though the example in Figure 2.6(c) illustrates the indirect pointer overwriting
the return address, it is possible to overwrite any interested pointers that have the control
flow information or that points to code that will be later executed.

2.3.2   Heap-based buffer overflows

Heap memory is dynamically allocated at runtime by the application. Similar to the case
with stack-based arrays (buffers), arrays in the heap can also be overwritten in almost all
implementations. The approach for overflowing is the same except that the heap grows
with the memory as opposed to stack that grows in the opposite direction. However neither return addresses nor frame pointers are stored on the heap and therefore an attacker
must use other means to gain control of the execution flow. Heap-based buffer overflows
have been understood and exploited in the computer underground for the last several

years, but the technique remains fairly mysterious. Vulnerabilities which causes stack-based buffer overflows have now been almost captured to extinction. However, heap-base buffer overflow vulnerabilities are still common. Heap-based buffer overflows can be divided into two categories: one comprises attacks where buffer overflow directly alters the content of an adjoining memory segment [173, 238], and the other incorporates misuses of management information used by memory manager such as malloc and free [5, 132]. Almost all the memory manager implementations share the convention of storing management information within the heap space itself and the aim of the attacks is to alter this information such that it will allow consequent inconsistent memory overwrites. Inconsistent memory overwrites could be used to change the control flow information such as return address as first illustrated in [71]. As classified under the first category, a simple way of exploiting a buffer overflow located on the heap is by overwriting heap-stored function pointers those are located after the buffer that is being overwritten [173]. But, as function pointers are not present in most of the instances, and therefore other mechanisms are used to gain control. Some of the other techniques are by overwriting a heap-allocated objects virtual pointer and pointing it to an intruder-introduced virtual function table [238]. When the vulnerable application attempts to execute one of these virtual methods, it will be redirected to execute the injected code to which the attacker-controlled pointer refers. As classified under the second category, overwriting the memory management information that is generally associated with a heap memory segment could be a general approach of attempting to exploit a heap-based buffer overflow, when function pointers or virtual function pointers are not present in the heap [5, 71, 132, 215]. Figure 2.7 depicts how a heap segment is managed by the memory manager (specific details on memory manager in this section are about GNU Lib Cs dynamic memory manager, *dlmalloc* implemented by Doug Lea [154]). Information of each heap segment (the sizes of the current and previous heap segments and the next and previous heap segment pointers - if its a freed segment) are stored in the heap segment header. The user data for a segment follows the header for that segment. When a heap segment is freed, part of its user data space is used to store a couple of pointers (as in Figure 2.7(a)), which are used to form a chain of freed heap segments.

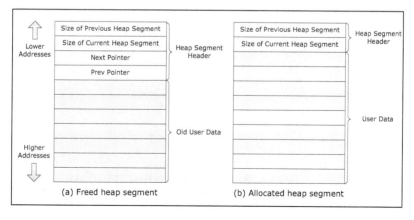

Figure 2.7: *Heap Segment Header and User Data*

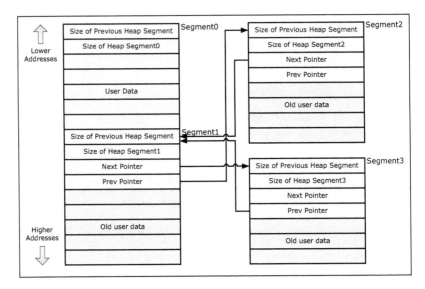

Figure 2.8: *Heap with used and free heap-segments*

Figure 2.8 depicts four heap-segments (*Segment0-Segment3*) in which one of the segments is allocated (*Segment0*) and all others are freed (allocated and freed using *malloc()* and *free()* or similar). As depicted in Figure 2.8, freed heap-segments are maintained in a memory segment chain (a doubly linked list of free segments) connected by the data structure as shown in Figure 2.9. As illustrated in Figure 2.8, when a segment is freed the memory normally used for data is used to store a forward and a backward pointer.

```
struct malloc_chunk{
        size_t prev_foot;          /* Size of prev chunk (if free) */
        size_t head;               /* Size of inuse bits          */
        struct malloc_chunk* next; /* double links (prev, next).. */
        struct malloc_chunk* prev; /* used only if free           */
};
```

Figure 2.9: *Data Structure of a Node in Freed Heap Segment Chain*

In Figure 2.9, *Segment2* is the first freed segment in the chain: its next pointer points to *Segment1*. *Segment1* is the next segment with its previous pointer pointing to *Segment2* and next pointer pointing to *Segment3*. *Segment3* is the last freed segment in our example chain with its previous pointer pointing to *Segment1*. A property of the memory manager to note here is that when two freed heap-segments are bordered, they will be merged into a single larger freed heap-segment. This merging operation is performed by a C macro called *UNLINK()*. When the memory manager at a later stage requests a heap-segment of the same size as one of the segments in the freed chain, then the first segment that matches the requirement will be removed from the chain and allocated to the program that requests the dynamic memory.

```
#define UNLINK(SEGMENT, PRV, NXT) {  \
      NXT         = SEGMENT → next; \ /* store pointers to  -  */
      PRV         = SEGMENT → prev; \ /* temporary variables */
      NXT → prev  = PRV;            \ /* chg next node's prev*/
      PRV → NXT   = NXT;            \ /* chg prev node's next*/
}
```

Figure 2.10: *Unlink Macro: Removing a Heap Segment from the Chain*

Figure 2.10 depicts the *UNLINK()* macro that will be used for the merge operation described in the previous paragraph. In Figure 2.10, *SEGMENT* is the heap-segment that is going to be allocated to the program and removed from the chain and *PRV* and *NXT* are temporary variables. The first and the second lines of the *UNLINK* macro (Figure 2.10) store the previous and next pointers of the heap-segment into the temporary variables. The third line de-references the previous pointer of the next heap-segment and set its value to the address of the heap-segment previous to *SEGMENT*. Similarly, the fourth line de-references the next pointer of the previous heap-segment and set its value to the address of the heap-segment next to *SEGMENT*.

Figure 2.11 shows what could happen if a buffer stored in heap-segment *Segment0* is overwritten by an attacker and the memory management information of *Segment1* is overwritten. In this example, the next pointer of *Segment1* is set to point to 12 bytes (3 x *sizeof*(size_t), a value of type *size_t* occupies 4 bytes) before the return address of the current function. The previous pointer of *Segment1* is set to point to the injected code in the buffer. When *Segment0* is later freed, it has to be merged with *Segment1* into a larger heap-segment because *Segment0* and *Segment1* are next to each other (bordered) in the memory. This is when the macro depicted in Figure 2.10 is called on heap-segment *Segment1*. Since the next pointer of *Segment1* is now pointing to 12 bytes before the return address, the return address will be overwritten with the previous pointer of *Segment1*. If the previous pointer of *Segment1* is set to point to (by the overflow) the injected malicious code, an attacker could gain access to the program execution flow. Even though this example is illustrated with a return address overwriting, the control flow of a program can be affected by overwriting other pointers.

2.3.3 Exploiting double-free vulnerability

De-referencing a freed memory location is not checked in most of the compiler implementations causing a problem in what is called dangling pointer references. A memory

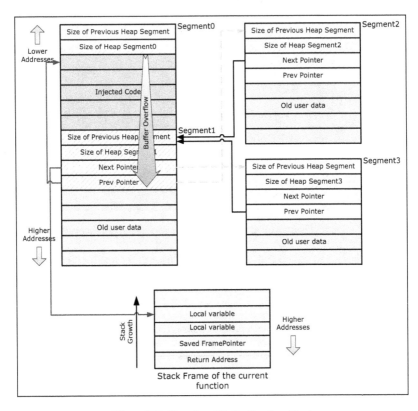

Figure 2.11: *Heap-based Buffer Overflow Attack*

allocation could be reversed either by explicitly calling function *free()* or by function epilogues generated by the compiler. Generally, de-referencing of a freed memory will cause unexpected program behavior or a program crash. But, double-free vulnerability is exploited not by using the freed memory, but by freeing it once again when it is already freed. A double-free vulnerability could allow an unauthenticated, remote attacker with read-only access to execute arbitrary code, alter program operation, read sensitive information, or cause denial of service [74].

Figure 2.12 depicts three freed heap-segments in a doubly link list chain as explained in Section 2.3.2. It is worth noting that when a heap memory request is processed by the memory manager, the first freed heap-segment that matches the size of the requested memory is unlinked and handed over to the program as explained in Figure 2.10.

Figure 2.13 depicts a segment of the *FRONTLINK* macro, which will be used by the memory manager when a heap-segment is freed and is to be linked to the linked list (this particular part of the code deals with inserting a node in between other nodes). In Figure 2.13, *P* is the pointer to the segment that needs to be linked to the chain; S is the

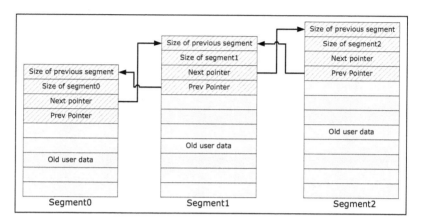

Figure 2.12: *List of Three freed Heap-Segments*

size of the freed heap-segment; *IDX* is the index to the segment chain; and *PRV* and *NXT* are two temporary variables of pointer to heap-segment type.

```
#define FRONTLINK( P, S, IDX, PRV, NXT) {          \
    ...                                             \
    PRV    = null;                                  \
    NXT    = start_of_bin(IDX);                     \ /* set NXT to front of list */
    while (NXT != PRV && S < size(NXT)){            \ /* search the right -      */
            NXT    = NXT → nxt;                     \ /* - location to place -   */
    }                                               \ /* - segment P of size S   */
    PRV             = NXT → prev;                   \ /* use PRV as temp to -    */
    ...                                             \ /* - store NXT→prev        */
    NXT → prev  = PRV → nxt = P;                    \ /* Insert P                */
}
```

Figure 2.13: *GNU Lib C's Frontlink Macro*

When a heap-segment is freed, the *FRONTLINK* macro searches the linked list to place the segment at the right location, so that it makes the smallest-first searching of the heap-segment allocation easier. The vulnerability of the *FRONTLINK* macro is that, if for example *Segment1* would be freed twice and therefore *FRONTLINK* is called upon *Segment1* again, the linked list will end up in a situation as depicted in Figure 2.14. That is, the *next* and the *previous* pointers of *Segment1* will point to itself. Now, assume that a heap-segment of the same size as *Segment1* is requested by a program and therefore the *UNLINK* macro (Figure 2.10) is called upon the linked list. If we substitute our values in the *UNLINK* macro:

```
NXT = Segment1 -> next = Segment1
```

```
PRV = Segment1 -> prev = Segment1
NXT -> prev = PRV = Segment1
PRV -> next = NXT = Segment1
```

Therefore, while the memory manager assumes that *Segment1* is allocated to the program, *Segment1* will remain in the same linked list as depicted in Figure 2.15.

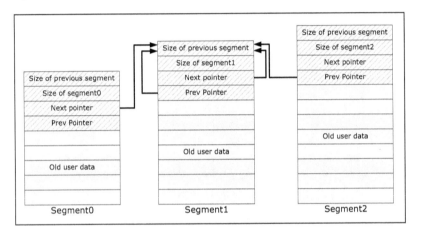

Figure 2.14: *List of freed Heap-Segment with* Segment1 *freed Twice*

Since the memory manager assumes that the *Segment1* has been unlinked and is free to be used by the program, as depicted by Figure 2.15, the program will be able to write its data anywhere under the user data portion of the heap-segment.

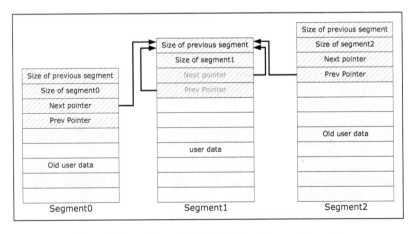

Figure 2.15: *Reallocated Double freed Heap-Segment (Segment1)*

Now an attacker may use the same vulnerability to exploit the memory as of heap-based buffer overflow (Figure 2.11) to overwrite the next and the previous pointers of *Segment1* with return addresses of any other control flow pointers to gain control over the execution flow of the program as described in heap-based buffer overflow attacks.

2.3.4 Integer errors

Integer errors are not exploitable vulnerabilities themselves, but could be used as a gateway to exploit one of the other vulnerabilities we discussed in subsections 2.3.1, 2.3.2 or 2.3.3. Integer errors could broadly be divided into two categories: Integer overflows and Integer signed-ness bugs [28, 326]. An integer overflow occurs when the value an integer has to hold grows beyond the maximum size of the value it could hold. ISO C99 [292] says that an unsigned integer that overflows is reduced by performing modulo with the number that is one greater than the largest value that could be represented by that type. Later in this section we will see how this same fact could be used to exploit an integer overflow for an attacker's benefit. Signed-ness bugs are the cause of misjudging either a singed value as unsigned or vice versa.

Integer overflows could be separated into width-ness overflows and arithmetic overflows. A width-ness overflow occurs when an attempt to store s a value in a variable which is too small to hold the variable occurs. When the value to be stored is from an arithmetic operation which overflows, any part of an application program that uses the result could be misled, and these types of overflows are called arithmetic overflows.

Figure 2.17 depicts a trivial example on exploiting width-ness overflow vulnerability. Type *unsigned short* is a 16 bits *unsigned integer* type and therefore the variable *len* could hold a maximum value of 65535 ($2^{16} - 1$). Storing a value that is greater than 65535 in *len* will result in reduced modulo 65356. For example, an attempt to store 65370 in *len* will result in storing 14 (65370 modulo 65356) in *len*. Running the code in Figure 2.17 with different values result as in Figure 2.16.

```
[roshanr@pera ~]$ ./widthness 20 "Hello World"          [attempt-1]
len = 20
Hello World
[roshanr@pera ~]$ ./widthness 120 "Hello World"         [attempt-2]
You can't do this!!
[roshanr@pera ~]$ ./widthness 65570 "Hello World"        [attempt-3]
len = 34
Segmentation fault
```

Figure 2.16: *Runtime Example for Exploiting Width-ness Overflow*

Even though the 'weak-check' in Figure 2.17 has detected a potential buffer overflow in *attempt-2* (in Figure 2.16), it did not detect it in *attempt-3* (in Figure 2.16) which caused a segmentation fault. While the code explained in Figure 2.17 may not appear exactly as shown in any real code, this serves an example for width-ness overflow.

Because of this width-ness overflow vulnerability, it is possible to bypass the bound checks as featured in *[weak-check]* in Figure 2.17 and potentially overflow the buffer.

```
/* An example of exploiting width-ness overflow
   based on blexim - <blexim@hush.com>        */

#include <stdio.h>
#include <string.h>
#include <stdlib.h>

int main(int argc, char **argv)
{
     unsigned short len;    /* maximum value 'len' could hold is 65535 (2^16-1) */
     char buf[100];

     len = atoi(argv[1]);
     if (len > 100){         /* check for buffer (buf) overflow */ [weak-check]
          printf("You can't do this!! \n");
          return EXIT_FAILURE;
     }

     printf("len = %d\n", len);
     strncpy(buf, argv[2], atoi(argv[1])); /* copy argv[2] to buf */

     printf("%s\n", buf);
     return EXIT_SUCCESS;
}
```

Figure 2.17: *A Simple Source code for Exploiting Width-ness Overflow*

Figure 2.18 depicts an example of exploiting arithmetic overflow due to its lack of verifying the value of the *length* parameter. The multiplication at (1) in Figure 2.18 depends on the value of *length* and one could overflow the variable *i* by carefully crafting the value of *length*. By choosing the value of *length* that is high enough to overflow *i*, it is possible to force the size of *buf* to be smaller than the size that is specified by *length*. Allowing an attacker to reduce the size of *buf*, will also allow the loop at (2) to write past the end of *buf* array and therefore ending up in a heap-based buffer overflow.

Integer signed-ness errors occur when a mix-up between signed and unsigned values take place. For example, when the programmer defines an integer, the system assumes it as a signed integer except when the variable is defined as unsigned. A signed negative value may pass a maximum size test that is designed for an unsigned value. Later when the signed value is cast into an unsigned value, this will possibly cause a buffer overflow if it is used with a memory copy operation.

Figure 2.19 depicts a piece of code where an integer signed-ness bug is present. The problem in this code is that the programmer fails to consider the fact that *memcpy* function's third argument (*length*) is an unsigned integer. The bound check at (1) in Figure 2.19 does the check assuming that the *length* is an integer (i.e. signed integer) and therefore it is possible to pass the check by passing a negative value length. At (2) when *length* is

```
/* An example of exploiting arithmetic overflow
    based on blexim - <blexim@hush.com>      */

#include <stdio.h>
#include <stdlib.h>

unsigned short arithoverflow(unsigned short *buffer, unsigned short length)
{
      unsigned short *buf, i;

      i = length * sizeof(unsigned short);  /* arithmetic overflow (1) */

      printf("i is %d and length is %d \n",i, length);

      if ((buf  = (unsigned short *) malloc(i)) == NULL)
            return EXIT_FAILURE;

      for(i = 0; i < length; i++)
            buf[i] = buffer[i];                    /* buffer overflow (2) */

      return EXIT_SUCCESS;
}
```

Figure 2.18: *A Simple Example for Exploiting Arithmetic Overflow*

passed as an argument to *memcpy* it will be interpreted as an unsigned integer (a very large number) and will cause the array *buf* to overflow. Overwriting a buffer is advantageous to an attacker and it could be exploited as explained in sections 2.3.1 and 2.3.2.

2.3.5   Exploiting format string vulnerabilities

A format function is a special function that takes a variable number of arguments, and expects one of which is a format string argument that controls the function. A format string is an ASCII string that contains text and format parameters. This format string is used to specify how the format function will access its other arguments that are given to the same function. Format functions are used in almost all C programs to output information, to print error messages, to prompt for user information, or to process strings. All the characters in a format string, except the format parameters starting with % character, are copied across to the output stream. The % character will be followed by format specifiers that will manage the way the output is generated. The arguments required by the format specifiers are expected to be found on the stack.

Figure 2.20 depicts an example for a format string function call (*printf*). Figure 2.20(a) depicts the function call itself, and Figure 2.20(b) and 2.20(c) depict the stack for *printf* function and an explanation of the stack elements respectively. When *printf* is called, it

```c
/* An example of exploiting integer signed-ness bug
   based on blexim - <blexim@hush.com>      */

#include <stdio.h>
#include <stdlib.h>
#include <string.h>

int verified_memcopy(char *input, int length)
{
        char buf[512];

        if (length > sizeof(buf))         /* signed int comparison    (1) */
                return EXIT_FAILURE;

        memcpy(buf, input, length);       /* length is cast to unsigned (2) */

        ...

        return EXIT_SUCCESS;
}
```

Figure 2.19: *Example for Exploiting Integer Signed-ness Error*

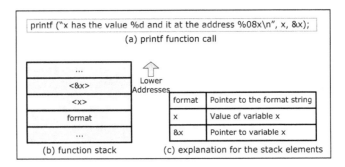

Figure 2.20: *Example stack for* printf

will process (parse) the format string (*format*) by handling one character at a time. When the character is not the % character, it is copied to the output. Each format parameter starting with the % character (expect %%, which represents printing % character itself) will be related to the data in the stack. In Figure 2.20(a), *%d* and *%08x* in the format string expects an integer and an address from the stack respectively. They are represented by *x* and *&x* in the stack in Figure 2.20(b). The data in the stack is popped and copied to the output according to the format string.

A format string vulnerability occurs when a format function as defined above, allows its user to input the format string. An attacker who is able to specify the format string to a format function will be able to control what this specific function pops from the stack and could make the program write to arbitrary memory locations of choice [256].

The elementary attacking technique using format string vulnerabilities is by supplying format strings to make a program crash. By crashing a program, it is possible to conduct denial of service attack or in some cases crashing a process daemon may dump cores which will reveal sensitive information. Passing the format parameter %s to a format function (example: *printf(%s)*) will display memory from an address that is supplied on the stack. Therefore passing more than few %s format parameters (example: *printf(%s%s%s%s%s%s%s%s%s%s%s%s%s%s)*) to a function will have a high chance of reading an illegal address and therefore causing a process crash that may lead to a core dump.

It is possible to use format functions to display useful control flow information from the stack by crafting the format strings carefully. Later this information could be used to find the right offsets of critical data (for example return address) for read exploitation or to reconstruct the stack frame of the target process. Figure 2.21 depicts an example piece of code that shows how a format function could be used to display the stack of that function. Since the *printf* function is called with %x argument, it will print the word from the top of the stack. Therefore, by using multiple emph%x arguments, it is possible to retrieve multiple number of parameters from the stack and display them as 8-digit padded hexadecimal numbers by using the format string %08x. Depending on the size of the format string buffer it is possible to retrieve all of the stack using this technique.

Running the code depicted in Figure 2.21 gives us an output as depicted in Figure 2.22. Since the format string had ten %08x formatting parameters we get the top ten words from the stack.

The %s format parameter of a format function uses an address (by reference) as a stack parameter (stack supplied address) and displays memory word from this address. Therefore it is possible to use *printf(%s)* to display data from a memory location pointed to by the address supplied into the stack by the same format function. Displaying arbitrary memory locations could be used to peek in memory locations for many reasons (for example, reconstructing process binary of a remote process [45]).

The %n format parameter is used to write into a variable of our choice, the number of bytes already printed. The address of the variable (integer pointer) where the number of bytes has to be written is placed as a parameter onto the stack. By exploiting this format parameter, it is possible to pick up an arbitrary address and write a value into it.

Figure 2.23 depicts an example for arbitrary memory overwriting using a vulnerable format function, and Figure 2.24 depicts how the vulnerability in the code depicted is exploited. The first attempt to execute the application *'formatstring'* as shown in Figure 2.24 is a genuine execution, and the second is vulnerable. From the first execution, the pointer to the variable x (*0x7fff7f58*) is identified and in the second run the pointer is used as a format parameter among the other parameters to change the value of variable x. There is the %n format parameter, which writes the number of bytes already printed, into a variable of our choice. The address of the variable is given to the format function by placing an integer pointer as parameter onto the stack. The fact that the %n parameter will write

```
/* An example of viewing the stack of a format function */

#include<stdio.h>
#include<stdlib.h>

int add(char *format, int x, int y)
{
        int z;
        z = x+y;
        printf(format, z); /* format function */
        return z;
}

int main()
{
        int x;

        x = add("The addition of 5 and 6 is %d\n",5,6);  /* genuine call */

        /* fake call */
        x = add("The stack has %08x.%08x.%08x.%08x.%08x.%08x.
                %08x.%08x.%08x.%08x\n",5,6);

        return EXIT_SUCCESS;
}
```

Figure 2.21: *Displaying the Stack of a Format Function*

```
The addition of 5 and 6 is 11
The stack has
0000000b.bfffd8b8.0804828d.b7fcee70.0000000b.bfffd8d8.080483f5.08048560.00
000005.00000006
```

Figure 2.22: *Output of Running the Code in Figure 2.21*

the number of bytes already printed into the address on the stack of our choice, is used to overwrite the value of x in this example. Note that, it is possible to write numbers which are much larger than the actual size of the buffer (100 in this example) into variable x. Therefore, writing an interesting value of an attacker's choice into a variable such as x is more than reality due to format string vulnerabilities.

A recent survey from *Microsoft Research* [213] shows real life examples for security threats beyond stack based buffer overflow attacks. The survey targets arc injection, pointer subterfuge, and heap smashing as its subject. Readers are referred to this survey for further details.

```
/* Arbitrary memory overwriting using a format function.
   An example based on Tim Newsham, Guardent, Inc. */

#include<stdio.h>
#include<stdlib.h>
#include<string.h>

int main(int argc, char **argv)
{
    char buffer[100];
    int x;

    if(argc != 2)
      return EXIT_FAILURE;

    x = 10;
    snprintf(buffer, sizeof(buffer), argv[1]); /* format function, uses   */
                                               /* argv[1] as format string */
    printf("buffer[0-%d] = %s \n", strlen(buffer), buffer);
    printf("x is %d and it is @ %p\n",x,&x);

    return EXIT_SUCCESS;
}
```

Figure 2.23: *Example for Arbitrary Memory overwriting with a Format Function*

$./formatstring "hello world"
buffer[0-11] = hello world
x is 10 and it is @ 0x7fff7f58

$./formatstring "\x58\x7f\xff\x7f%.520d%n"
buffer[0-99] = \x58\x7f\xff\x7f0000000000.....000
x is 524 and it is @ 0x7fff7f58

Figure 2.24: *Running the code in Figure 2.23*

2.4   Reliability Issues

The reliability of a computer system is a function that depends on time and is defined by the probability, that the lifetime of the system that exceeds a predefined time. A system failure is defined by the behavior of the system that diverges from that prescribed by the user. Either a failure of a system component or an erroneous design is the cause for fallacious system state transition, which will lead to a system failure. An error may or may not lead to a failure will depend on the system configuration or on the system operation.

Identifying the effects of faults and the faults themselves in an embedded processor is a topic of unceasing research. The experimental studies conducted by Siewiorek et al. [267, 268] reveals that more than 90% of system faults are caused by non permanent errors, that is transient faults. Transient faults are caused mainly by the following factors.

1. Ionization by cosmic rays or alpha particles (from packaging material);

2. Power fluctuation; and

3. Electro-magnetic interference.

Furthermore, a very small number of irregular faults occur due to design faults. These errors are generally known as logic soft errors or single event upsets (SEU) [261, 273]. These errors cause unintended transitions of logic states in a processor which if not detected, results in undesirable outcomes.

2.4.1   Single Event Upsets

Soft errors are errors caused due to random events resulting in corruption of data but are neither repeating nor permanent. Soft errors are increasingly becoming responsible for system faults in computers and are becoming one of the significant causes for unreliability. Even though soft errors have been seen as errors in memory modules in the past, soft error rate per chip of logic circuits have increased significantly in recently [189, 266]. Due to the demand for high performance and low cost embedded systems, reliability is rarely a priority in their design process. As the technology advances device scaling, reduction in feature size, and decreasing transistor voltage level are causing soft errors, which are capable of provoking hardware faults. Generally software errors (due to software vulnerabilities) are the known cause for unavailability, but recently due to the faster, denser, larger, complex processors, memories and caches soft errors are tending to beat the software errors in causing unavailability.

Soft errors impact different resources, such as either the transient error in logic, or a bit upset in a register which changes the value in register used by the software. In this era, system memory is more vulnerable to soft errors compared to the processor cache and system bus with parity check protection. Error Correction Code (ECC) is used in system memory to protect from data or code error. Because of the difficulty of field study of soft errors, often fault injection methods are used to measure the impact of soft errors on systems. Software fault injection methods are easy to implement and do not require experimental equipment to perform fault injection as opposed to hardware fault injection experiments [122]. In fault injection analysis, overwritten errors (errornous bits overwritten by a memory store) are ignored as they do not affect the processor state. Single bit errors and register file upsets, and message passing interface (MPI) message payload corruption have been simulated using fault injection methods in [62]. The results indicate that some applications are very sensitive even to single bit errors.

Control Flow	Error in the sequence of instructions executed by the main processor in the system
Data	Changes in the data values due to a transient fault
Control Signal	Changes in the control signals and no changes either in the control flow or data
Uncategorized	Changes in processor signals which could not be categorized under the above three

Figure 2.25: *Categories of Transient Faults*

Another recent experiment has been reported in [178] where the authors have used an IA32 platform to simulate memory errors with watch points. The results show that the soft error has clear impact in system software. The findings by [178] are:

1. majority of the memory errors require only partial recovery as the memory bits in concern in general are overwritten by memory stores before the erroneous bits are used;

2. registers and input/output buffers have a high probability of being affected by single bit errors and they have high impact on the result;

3. although the application heap has higher activation and susceptibility for errors compared to static data, its impact in user program is low;

2.4.2 Organization of Transient Errors

In [138], the authors has classified transient errors into four categories as depicted in Figure 2.25. The error classification in [138] is done by performing physical fault injection analysis (explained in section 3.5) to MC6809E, a non-pipelined complex instruction set computer (CISC) based microprocessor. As depicted in Figure 2.25, the errors are classified under data errors, control flow errors (could either be temporary or permanent), control signal errors and uncategorized errors. The authors in [206] have extended the classification in [138] to include some other basic error categories such as address error, bus error, illegal opcode, etc. Simulation based fault injection methods were used with the register-level model of a RISC processor in VHDL. The authors of [206] have shown that the RISC processors are more prone to data errors than the CISC processors used in [138].

Another classification of transient faults is presented in [251]. This classification is performed based on the behaviors of control flow, memory access and address values. The categories are: (1) invalid program flow, (2) invalid op-code address, (3) accessing unused memory, (4) invalid read address, (5) invalid write address, (6) invalid opcode, and (7) non-existent memory. This study uses modified system bus signals to simulate the transient faults. Another similar classification is performed in [166] with same fault injection technique and additional external circuits for injecting faults. From these experiments we could see that the important error detections are performed on control flow, memory access behavior, and instruction code.

2.5   Summary of this chapter

In this chapter, a summary of known security and reliability problems in computing systems and how they are connected to embedded systems are shown. Further a detailed analysis of all the known code injection attacks which are the major security threat of this era is given. In the later sections a discussion of some common reliability problems due to transient faults, particularly due to single event upsets is given. This section is concluded by saying that we not only need error detection and recovery mechanisms for soft errors only in registers and memory but also for all other circuits in a processor.

Literature Review

... 'what is the use of repeating all that stuff, if you don't explain it as you go on? It's by far the most confusing thing I ever heard!' ...

— Lewis Carroll, *Alice's Adventures in Wonderland*

This chapter presents a detailed survey on the available solutions in the literature to the security and the reliability problems in computer systems. This survey spans from the high level software solutions to the system level hardware/architectural solutions.

3.1 About this Chapter

3.1.1 Objectives

This chapter has four primary goals and they are:

1. summarising all the known proposed solutions to security problems in computer systems in general, and countermeasures to code injection attacks in particular;

2. giving an abridged version of all the known reliability support solutions such as soft error detection, control flow checking, etc.

3. detailing known hardware based security and reliability frameworks and their insights; and

4. presenting an analysis of all the known fault injection analysis and their pros and cons.

3.1.2   Outline

The material presented in this book combines two elements: defining a hardware software framework for reliability and security monitoring at the granularity of micro instructions, applying this framework for real world problems in reliability and security. This chapter provides literature review on how these problems are handled by other researchers and how previous research projects have motivated the work of the thesis presented in this book. In addition, comparisons between the existing literature are presented here, so that it could highlight the uniqueness of the work presented in the latter part of this book.

3.1.3   Contributions

The major contribution of this chapter is to provide a literature survey on solutions to the most encountered security and reliability problems. However, comparisons among these solutions are also summarized in this chapter.

3.2   Hardware Based Security and Reliability Frameworks

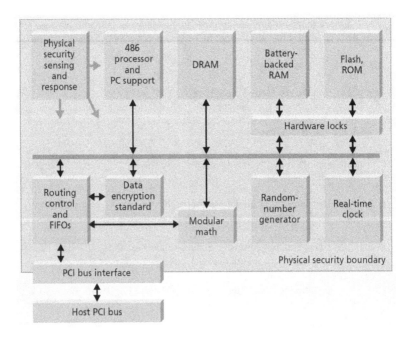

Figure 3.1: *Architectural Overview of IBM's 4758 Secure Coprocessor (taken from [78])*

Most of the earlier research on hardware assisted security and reliability framework techniques concentrated on the implemention of tamper resistance and cryptography systems.
In [99], the authors argue that security is one of the crucial supports needed by current
embedded processors. In [78], the authors have presented an IBM secure coprocessor
(IBM 4758) that provides physical tamper resistance and hardware support for cryptography. The paper claims that this work was initiated in the 1980s when the researchers
started looking at tamper responsive hardware to prevent software piracy. The hardware
architecture and the building blocks of the IBM 4758 is depicted in Figure 3.1. The
IBM 4758's hardware architecture is built around a 486 processor core, in which the coprocessor contains two types of battery-backed RAM, a random-number generator, and
persistent storage space for encrypted data in flash memory.

Untrusted external memory is one of the major causes for security and reliability problems. Therefore, verifying the memory integrity will be a vital technique when implementing secure and reliable processors that are resistant to either attacks or due to failed
hardware components. In [281], Suh et al. have proposed a memory integrity verification
technique that uses a hardware scheme to ensure the integrity of the untrusted external
memory component. In this secure computing model, the processor core is assumed to be
invulnerable to physical attacks and only the external memory is untrusted.

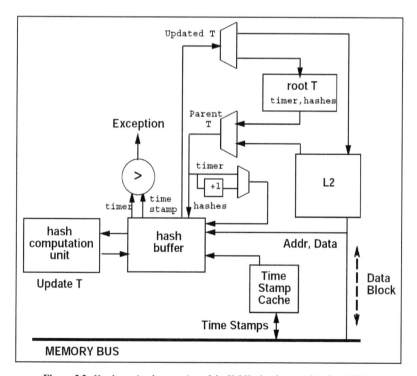

Figure 3.2: *Hardware implementation of the H-LHash scheme (taken from [281])*

The scheme proposed in [281] uses a very small amount of trusted on-chip storage and an integrity checking module is added to the hardware. Figure 3.2 depicts the hardware implementation details of the *H-L Hash* integrity checking scheme. The integrity check-ing module is added to the Level 2 (L2) cache and whenever a cache miss occurs the data from the memory is verified before being loaded to the trusted L2 cache. Furthermore, their scheme maintains incremental multiset hashes of all memory reads and writes which can be used for verification at a later stage.

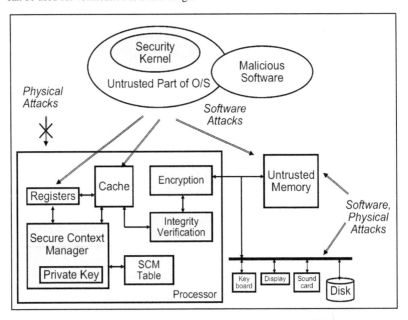

Figure 3.3: *AEGIS: Secure Computing Model (taken from [282])*

In [282], Suh et al. have proposed a single-chip processor, AEGIS that is secure against both physical and software attacks. Figure 3.3 depicts the secure computing model of the AEGIS technique. AEGIS provides its users with tamper-evident and authenticated environments in which any physical or software tampering by an adversary is guaran-teed to be detected. Furthermore, AEGIS also provides a private and an authenticated tamper-resistant environment, where the adversary is unable to obtain any information about software or data by tampering with, or otherwise observing, system operation [282]. As depicted in Figure 3.3, the AEGIS system is built around a processing subsystem with external memory and peripherals. Some of the applications where AEGIS could be used are: (1) grid computing, (2) secure mobile agents, (3) software licensing, and (4) digital rights management [282].

In [227], the Ravi et al. survey various tamper or attack techniques and explain approaches that have been proposed to design tamper-resistant embedded systems. Figure 3.4 depicts the specific objectives of tamper-resistant design approaches. Their approach starts from attack prevention, making an attack difficult. In case of a successful attack, there should

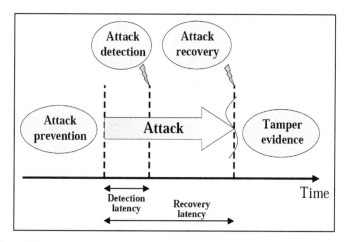

Figure 3.4: *Specific objectives of tamper-resistant design approaches (taken from [227])*

be a mechanism to detect this and to impose recovery mechanisms. Furthermore, there should be a mechanism to preseve attack logs for further analysis at a later stage. For a detailed analysis on this process, refer to [227].

In [192], Nakka et al. have proposed a framework called Reliability and Security Engine (RSE), which is an additional hardware unit sitting next to the core processor, that could be used to monitor different security and reliability properties. RSE performs security and reliability checking based on the application which runs on the main processor which is protected by it. RSE relies on compiler support to insert check instructions into the instruction stream. The monitoring hardware in RSE is external to the core processor and therefore needs a well defined interface (ports) within the core processor to communicate with RSE. Since RSE is a separate hardware unit, it needs self-checking mechanism to ensure that this additional hardware is reliable and not compromised. In [300], the author has implemented two of the RSE modules in hardware using VHDL and synthesized it onto an FGPA to evaluate area and clock period overheads.

In Figure 3.5, two different instruction execution scenarios, the synchronous and the asynchronous modes which are used by the RSE framework. Some other RSE modules which support either process hang or process crash are proposed in [193]. These modules are (1) Instruction Count Heartbeat (ICH), that detects process crashes and hangs, where instructions are not executed; (2) Sequential Code Hang Detector (SCHD), that detects process hangs due to illegal loop formation; and (3) Infinite Loop Hang Detector (ILHD), that detects infinite loop formation from proper loops.

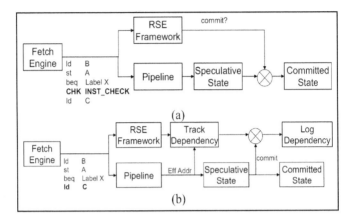

Figure 3.5: *RSE Instruction Execution Scenario (taken from [192])*

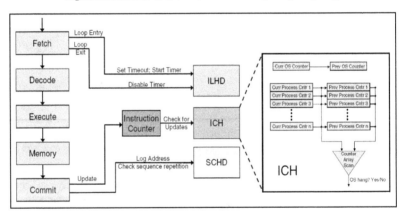

Figure 3.6: *Block Diagram of Processor with Process Crash/Hang Detection Modules (taken from [193])*

3.3   Defense Against Code Injection Attacks

There are several countermeasures proposed in the literature to defend against code injection attacks performed by exploiting common implementation vulnerabilities and they are discussed in this section. I have divided the countermeasures into nine groups based on (1) the system component where the proposed countermeasure is implemented; and (2) the techniques used for the countermeasures. Following are the nine groups discussed here:

1. Architecture based Countermeasures
2. Safe Languages
3. Static Code Analyzers

4. Dynamic Code Analyzers
5. Anomaly Detection Techniques
6. Sandboxing or Damage Containment Approaches
7. Compiler Support
8. Library Support
9. Operating System based Countermeasures

3.3.1   Architecture based Countermeasures

As return addresses of functions are the most attacked target of buffer overflows, there are
many hardware/architecture assisted countermeasures that aim to protect these addresses.
In 2002, Xu et al. [323] proposed two different countermeasures to protect against buffer
overflow attacks and they are:

1. the secure return address stack and,

2. divided control and data stack

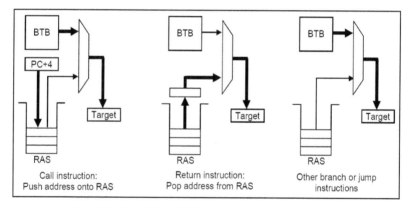

Figure 3.7: *Secure Return Address Stack (SRAS) (taken from [177])*

Secure return address stack (SRAS) is a proposal to extend, the already existing return address stack (RAS) used by the processors for return address prediction [270, 309]. When
a call instruction is fetched, these processors predict the return address to allow the processor to pre-fetch the next instructions when executing a return instruction. Xu et al.
propose a non-speculative RAS, instead of the speculative one, meaning that the RAS
will be updated in the commit stage, not in the fetch stage of the call instructions. This
will eventually cause a mismatch between the address from RAS and the return address
stored in the stack, when there is a buffer overflow that alters the stack and the return
address stored in the stack. Dividing the control and data stack is a rudimentary solution
to the problem of overflowing a buffer and overwriting the control flow information. The
control information (return address, etc.) are now stored in a different stack than that of
the other data (buffers, other local variables, function arguments, etc.). It is possible to

use software and hardware mechanisms to do this and Xu et al. have used a compiler implementation, where the function prologue and epilogue are modified to accomplish different stacks for control and data. Xu et al. have also proposed a hardware technique to implement this technique where they have to change the related instruction behavior along with minimal modification to the processor (adding special stack pointers to point to the new control stack).

Around the same period when Xu et al. proposed their SRAS, Lee et al. [155, 177] proposed a similar SRAS suggestion, as did Ozdoganoglu et al. [209]. Lee et al. have also proposed techniques to handle *setjmp* and *longjmp* calls such as (1) prevent from using these calls, (2) create extra instructions that allow the compiler to manually push and pop values from SRAS when *setjmp* and *longjmp* are called. They have also suggested allowing users to turn on and off the SRAS for a particular application. Ozdoganoglu et al. proposed a technique called *SmashGuard*. When there is a mismatch between the return addresses from SRAS and the regular stack *SmashGuard* which will search down the regular stack until it finds the matching value or until it reaches the end of it. The drawback in this technique is that the *SmashGuard* will stop at the first match for the return address while the real return address maybe at a location further down the stack, after a particular *longjmp*.

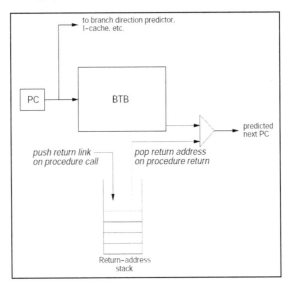

Figure 3.8: *The Basic Operation of a Return Address Stack (taken from [270])*

Shao et al. proposed a hardware assisted protection against buffer smashing attacks and function pointer attacks [263, 264]. The proposed hardware/software address protection (HSAP) consists of two components: the first uses hardware boundary checking to protect against buffer smashing attacks and the second uses function pointer XOR method to protect against function pointer attacks. The boundary check is performed against values inserted at compile time and protect against buffer overflows at memory write stage of the pipeline stages. Function pointer XORing is used to confuse the attacker and to make it extremely hard to change the function pointer to point to the malicious code. A randomly assigned key for each process will be used for XORing and therefore will be hard to perform function pointer smashing attacks.

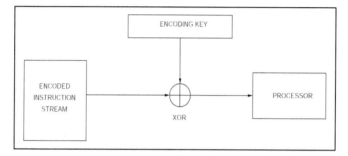

Figure 3.9: *Instruction-set Randomization (taken from [140])*

Randomizing instruction set of a processor makes it impossible for an attacker to guess the encrypted instructions and therefore prevents from injecting malicious code into the memory at runtime. Because each process will be encrypted with a different key at load time and decrypted before execution and the attacker is unable to guess the decryption key of the current process, even an attempted injection will result in a process crash due to wrong instructions being executed. Barrantes et al. use an x86 processor emulator to encrypt the instructions at load time and decrypt them before execution [21]. Encrypting the application at load-time will keep the original program unchanged and therefore could be encrypted with different keys, each time the program is executed. Kc et al. proposes a method to encrypt the program on the disk instead of at load-time [140] and including the key in the header of the program. Then at load time the key is stored in a special register and used for decryption before instruction execution. Storing the encrypted program on the disk opens more security threats and so does the keys in the header of the program. Even though randomizing instruction set seems to be a feasible solution, the performance overhead for decrypting each instruction at runtime is large and makes this approach impractical.

In [11], Arora et al. use an additional co-processor and hardware tables to perform software integrity checks. They identify program properties at different levels of granularity and store multiple control flow levels of data and checksums to perform software integrity monitoring. Figure 3.10 depicts an overview of their proposed framework and it includes three different checkers: (1) inter-procedural control flow checker; (2) intra-procedural control flow checker; and (3) instruction integrity checker. Tables with control flow data are loaded from the compile time information and they are verified at runtime. Therefore the method produces code which is not relocatable.

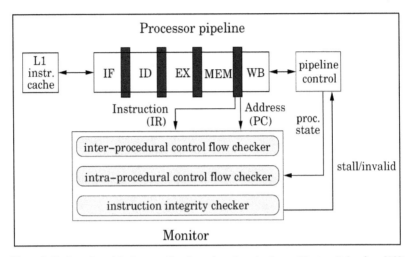

Figure 3.10: *Overview of the Proposed hardware-based monitoring architecture (taken from [111])*

3.3.2 Safe Languages

Safe languages such as Java and ML are capable of preventing some of the implementation vulnerabilities discussed here. However, every day programmers are using C and C++ to implement more and more low and high level applications and therefore the need for safe implementation of these languages exists. Languages close to C and C++ known as safe dialects of C and C++, use techniques such as restriction in memory management to prevent any implementation errors.

Cyclone is a safe dialect of C language[110, 128]. Cyclone statically analyses the C code and inserts dynamic checks at places where it cannot ensure that the code is safe. Cyclone further refuses to compile until more information is provided to prove that the program is safe. In Cyclone, uninitialized pointers are not allowed in the program and automatic tag injection is used to avoid format string vulnerabilities by verifying the type of arguments used in a function. Region based memory management schemes are used to prevent the dereferencing dangling pointers [111, 120]. Cyclone implementation has a heap, a stack and a dynamic memory region. The heap has no *'free'* operation, the stack is de-allocated on return of a function, and the dynamic segments are deallocated as a whole. These features prevent the vulnerabilities detailed in chapter 2. Cyclone restricts programmers by disallowing use of unsafe program constructs. Since software is built on top of existing libraries (for example the *GNU C Library* - libc), it is not feasible to use languages like Cyclone to write real applications in practice, as we need to rebuild libraries with Cyclone.

CCured [198, 200] is a source-to-source translator for C and is based on C Intermediate Language (CIL, [199]). Similar to Cyclone, CCured also does static analysis and inserts dynamic checks in the program if required, to prevent all memory safety violations. However, unlike Cyclone (where a programmer can decide which type of pointers are used in a specific place) the type of pointers to be used are determined by the CCured itself based on a static analysis. Some of the problems which arises on compatibility of CCured compiled code with non CCured compiled code are discussed and solutions are presented in [53].

In Vault [153, 182], the authors have proposed a method that is different from Cyclone on implementing a region-based memory management system. In Vault, the resource management protocols are enforced statically by the compiler by extending the type system and can be described by the programmer as described in [66]. Control-C [73, 147], which is designed for real-time control systems that runs on Low Level Virtual Machine (LLVM), is used to provide memory safety for unsafe C programs without annotations, garbage collection or run-time checks. This approach supports a subset of C language with the assumption of statically ensured memory safety.

3.3.3   Static Code Analyzers

Static Code Analyzers, analyses the software without actually executing programs built from that software [250]. In most cases the analysis is performed on the source code and in the other cases on some form of the object code. The term analyzer usually means that the analysis is performed by an automatic tool compared to human analysis which is done manually. The maturity of the analysis performed by these tools ranges from those that only consider the behavior of simple statements and declarations, to those that include the complete source code of a program in their analysis. The information collected by these analyzers could be used in a range of applications, starting from detecting coding errors to formal methods that mathematically prove program properties.

Although a thorough static analysis may reveal a significant number of vulnerabilities, all static code analyzers will generate a number of false positives (the amount of correct code that is incorrectly reported as being vulnerable), false negatives (vulnerable code that is not reported as vulnerable) or both [327]. Therefore, the effectiveness of a static code analyzer is determined by the ratio between the false positives and false negatives generated. Even though the analysis performed by static code analyzer is automatic, for some analyzers, the programmer needs to place annotations inside the source code (annotated static code analyzers). This increase the burden on programmer. The other type of static code analyzers are non-annotated, where analyzers themselves will infer what the programmer implied from the source code. Annotations make the life of the analyzer relatively easy as it has to do less or zero inferring about what the programmer intended. It is also advantages of having annotations as the analyzer could look for annotations in the code for analysis rather than analysing the whole code. Non-annotated static code analyzers generally attempt to confirm that the code is safe by forming a model of the execution environment at every step of the program code and asserting safety.

An annotated static code analyzer called *Splint* is demonstrated in [83, 151]. It is a lightweight tool, which uses annotations of specific properties to objects or pre- and post-conditions of functions. Splint could be used with a small effort to analyze software code for some of the common implementation errors. Figure 3.11 depicts such annotations for some selected C library functions. The annotations in Figure 3.11 were determined based on ISO C standard (ISO99).

```
char *strcpy (char *s1, char *s2)
   /*@requires maxSet(s1) >= maxRead(s2)@'
   /*@ensures maxRead(s1) == maxRead (s2)
          /\ result == s1@*/;

char *strncpy (char *s1, char *s2,
               size_t n)
   /*@requires maxSet(s1) >= n - 1@*/
   /*@ensures maxRead (s1) <= maxRead(s2)
          /\ maxRead (s1) <= (n - 1)
          /\ result == s1@*/;
```

Figure 3.11: *Examples for Annotated C Library Functions (taken from [151])*

In [75, 76], Dor et al. demonstrate, C String Static Verifier (CSSV), a tool that statically uncovers all string manipulation errors. CSSV is another annotated static code analyzer tool that requires the programmer to provide descriptions of pre-, post-conditions and the side-effects of functions. Figure 3.12 depicts a high level structure of CSSV. Pointer analysis is performed on the annotated source code to detect the pointers that point to the same base address. Based on this result, inter analysis [55] is performed and all the assertions are verified. A failure in one of the assertions is reported as a potential error. CSSV is extended in [269] with tracking of definite changes to a buffer as opposed to tracking all possible changes, reducing large number of reported false positives, etc.

Figure 3.12: *High Level Structure of CSSV (taken from [75])*

Figure 3.13 depicts the architecture of another annotated static code analyser tool [262], which is extended from the *cqual* [94] system, an extensible type qualifier framework. The C source code and the configuration files are parsed, producing an annotated Abstract Syntax Tree (AST). Cqual traverses the AST to generate a system (or database) of type constraints, which are solved on-line. Warnings are produced whenever an inconsistent constraint is generated. The analysis results are presented to the programmer in an emacs-based GUI, which interactively queries the constraint solver to help the user determine the cause of any error messages.

A compiler extension technique that falls under annotated static code analyzers is proposed in [13]. Under this framework the programmers could write specific rules as compiler extensions and the compiler at compile time verifies whether these rules are followed by the source code it compiles. The authors demonstrate this by looking for integer usages in untrusted source code and flag potential integer errors and array bound violations.

PREfix [38] and the extension of PREfix called PREfast [153] are two non-annotated static code analyzers, which build execution models of the analysed source code. This model represents all possible execution paths of an entire program. A full memory model is derived from the execution model and is used to detect potential inconsistencies between the expected and actual values. An example for PREfix path tracing is given in

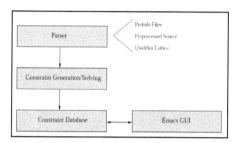

Figure 3.13: *The architecture of the cqual system (taken from [262])*

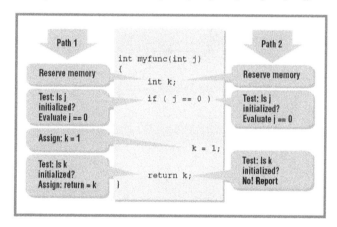

Figure 3.14: *The PREfix tool. PREfix traces two paths through this function, produces a model, and analyzes it for errors (taken from [153])*

Figure 3.14. It is worth noting that the PREfix tool resembles a dynamic code analyzer, called Purity [117], which is explained in the following section.

Figure 3.15: *The architecture of the buffer overflow detection prototype (taken from [305])*

A non-annotated static code analyzer is described in [305] that finds potential buffer overflow vulnerabilities in C code. The static analysis is formulated for the detection of buffer overflows as an integer range analysis problem. Each string in a program is eventually represented by two integers and they are (1) the number of bytes allocated for a particular string and (2) the number of bytes currently in use related to that string. The language is designed to constrain specific string operations based on this integer range. When the constraints are generated, the possible execution paths are analyzed and places where the execution violates these constraints are reported as potential overflow points. The authors

in [305] have implemented their design in a prototype (as depicted in Figure 3.15) to find
new remotely-exploitable vulnerabilities in a large, widely deployed software package.

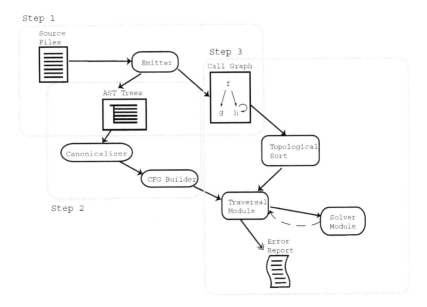

Figure 3.16: *An Overview of the ARCHER Analysis (taken from [320])*

Another non-annotated static code analyzer, ARray CHeckER (ARCHER) is proposed
in [320]. It is implemented to look for memory access violations via evaluating array accesses and pointer dereferences using a constraint solver. Accesses that violate constraints
are reported as errors. Figure 3.16 depicts a flow-chart of all the three steps involved in
the analysis of a program using ARCHER. In *Step 1*, the C source code is parsed with
GNU C Compiler (GCC) and the serialized abstract syntax tree (AST) is written back to
files. In *Step 2*, the ASTs are made into canonical form, before building a control flow
graph (CFG) from the ASTs. In *Step 3*, the static code analysis is performed by traversing
the CFG and inspecting each function for potential errors. Different techniques are used
for detecting errors depending on the type of the statements, such as boolean conditions,
memory accesses, etc. The primary limitation of ARCHER is that it only operates on arrays and pointers in programs and does not comprehend string operations. Furthermore,
memory for which no size can be determined is also not checked under the ARCHER
framework.

Errors in parallel programs are detected in [243, 244] by using a non-annotated static
code analyzer. The framework proposed by the authors performs the symbolic bounds
analysis of pointers, array indices and accessed memory regions. Figure 3.17 depicts the
general structure of the code analyzer. The analyzer consists of a pointer and read-write
sets analysis and a symbolic analysis. When the analysis phases are done, the information
from these phases are used to perform the bound checks.

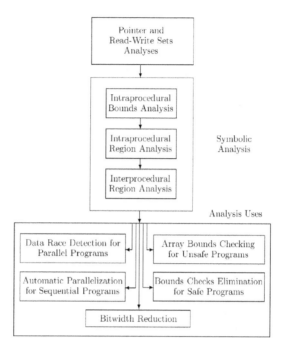

Figure 3.17: *The Structure of the Compiler Proposed in [244] (taken from [244])*

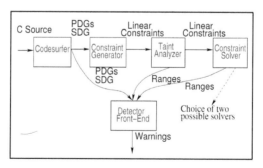

Figure 3.18: *Overall Architecture of the Buffer Overrun Tool (taken from [98])*

In [98], Ganapathy et al. demonstrate a technique for statically analyzing C source code by modeling C string manipulations as a linear program. Under this technique a buffer is modelled by four integers representing the maximum and minimum amount of bytes allocated and used. Based on this model, the authors have built a prototype and used it to identify several vulnerabilities in popular security critical applications. Figure 3.18 depicts the overall structure of the tool. The tool consists of five components and they are:

1. *Code Surfer* is the code understanding component;

2. *Constraint Generator* uses information from code surfer to generate constraints;

3. *Taint Analyzer* makes the constraints manageable by the constraint resolver;

4. *Constraint Resolver* uses linear programming to solve constraints that remain after taint analysis ; and

5. *Detector front-end GUI* is used to help the user examine potential buffer overflows.

In [301], Viega et al. aim to produce a lightweight static code analyzer, so that they could implement this as an editor extension. They have proposed ITS4 that attempts to find unsafe C and C++ code. ITS4 does only lexical analysis as opposed to most of the other techniques which parses the code (semantic analysis) and will report vulnerable code by matching the source code with a vulnerability database. When it has identified possible vulnerable functions, it will examine them further and determine if they are to be reported and with what level of severity. *Flawfinder* [310] is also a tool similar to ITS4 that uses lexical analysis to detect vulnerable functions. RATS [257], a tool similar to Flawfinder, is capable of analysing programming languages such as Perl, PHP and Python in addition to C and C++.

3.3.4   Dynamic Code Analyzers

In dynamic code analysis, the source code is instrumented at compile time and then test runs are performed to detect vulnerabilities. Even though performing dynamic code analysis is more accurate than static analysis (more information of the execution is available at runtime compared to compile-time), the dynamic code checking might miss some errors, as they may not fall on the execution path while being analyzed. Some well known dynamic code analyzers are discussed in this section.

Purify [117], which is later known as IBM® Rational® PurifyPlus™ [25, 41] is a software testing tool and debugger, designed for advanced runtime and memory management error detection. Even though Purify family is not primarily designed for security, it detects memory leakages and access errors which are some of the security vulnerabilities discussed in this book. According to IBM's website, PurifyPlus's runtime check includes four basic functions and they are (1) memory corruption detection; (2) memory leak detection; (3) application performance profiling; and (4) code coverage analysis.

In [118], Haugh et al. introduced a dynamic buffer overflow detection tool for C programs, STOBO, which does compile time instrumentation to the application so that the runtime states of the applications are not affected. Based on whether the source and/or the destination of buffers are statically and/or dynamically allocated, different types of warnings are raised, when either the destination is smaller than the source or no adequate size checking is available for buffers.

Another dynamic analyzer is proposed in [152] to detect invalid array accesses through software instrumentation and runtime checking. The applications are not only tested for

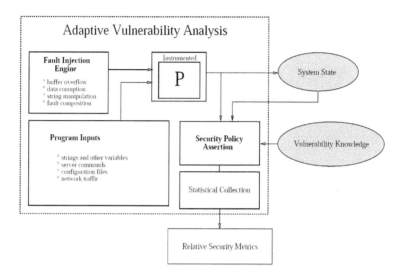

Figure 3.19: *An overview of the FIST tool (taken from [102])*

the default inputs, but also for many more inputs which generate a range of values for all the input variables. This will cover most of the undetected errors in other dynamic analyzers due to some error cases missing out, as they may not fall on the execution path.

In [90], program validation is done by performing a dynamic property-based testing. In this work the program execution is considered as a sequence of state changes. A tool called the *instrumenter* takes program specifications written in a specification language and modifies source code so that those changes of state relevant to the specifications will produce output to a state change file. Another tool, *the test execution monitor* takes that state file and the program specifications, and reports any violations. This enables a tester to determine when the program fails to meet its specifications. This is not a formal verification, but a dynamic code analysis and testing technique.

In [101], Ghosh et al. introduce Fault Injection Security Tool (FIST), a dynamic code analyzer that detects software vulnerabilities with the help of software fault injection. A wide range of errors such as changing boolean values, exploiting buffer overflows, etc. are injected during execution time of applications and violations of security policy are observed from the applications' responses to the inputs. The fault injection technique is capable of simulating stack based buffer overflow attack, which will overwrite the return address of the function . However, the testers has to manually identify the buffers to overflow and instrument them in [101]. An automatic process for identifying buffers to test is proposed in [102], where the source code is parsed, analysed, and instrumented at places that might be vulnerable to attacks.

An overview of the FIST is shown in Figure 3.19. A program, P, is instrumented with fault injection functions and assertions based on the vulnerability knowledge of the program. The program is executed using different inputs. The security policy is dynamically evalu-

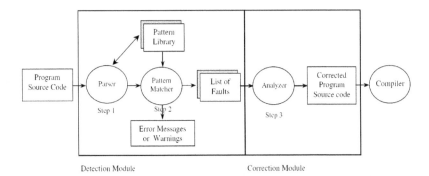

Figure 3.20: *An Overview of Precompiled Fault Detection (taken from [65])*

ated using program and system states. If a security policy assertion is violated during the dynamic analysis at runtime, the specific input and fault injection function that triggered the violation is identified.

In [65], Deeprasertkul et al. have introduced a Precompiled Fault Detection (PFD) technique. This is a technique that automatically detects and corrects programming errors, which are caused by programmers' ignorance and cannot be detected by a compiler, in the source code prior to compilation time. Even though the PFD technique proposed in [65] is capable of detecting errors in C programs, the authors claim that this as a general technique and could be extended to work with other language applications in the future. Figure 3.20 depicts an overview of the PFD approach. In *Step 1*, the source code of the program is parsed to discover the program statements with potential faults. In *Step 2*, the *Pattern Matcher* considers the variables, function calls, etc. to match the pattern of declarations which are generated in *Step 1*. The *Pattern Matcher* also generates log files for the faults defined in PFD. In *Step 3*, most of the faults are corrected automatically by the *Analyzer*.

3.3.5 Anomaly Detection

Behavior-based anomaly detection compares a profile of all allowed application behavior to actual behavior of the application. Any deviation from the profile will raise a flag as a potential security attack [121]. This model is a positive security model as this model seeks only to identify all previously known good behaviors and decides that everything else is bad. Behavior anomaly detection has the potential to detect several type of attacks, which includes unknown and new attacks on an application code. Most of the time the execution of system calls is monitored and is recorded as an anomaly, if it does not correspond to one of the previously gathered patterns. A threshold value for the number of anomalies is decided a priori and when the threshold is reached, the anomaly can be reported to the system and subsequent action, such as terminating the program or declining a system call can be taken.

On the negative side, behavior anomaly detection can lead to a high rate of false positives. For instance, if some changes are made to the application after a behavior profile is created, behavior-based anomaly detection will wrongly identify access to these changes as potential attacks.

In [304], Wagner et al. have proposed four different means to detect anomalous nature of applications through static information collection of system calls. In the first approach, all the system calls an application may execute are stored statically and are verified with runtime system calls. A runtime system call that is not in the stored set is identified as an anomaly. The second approach that has been proposed in [304] is very similar to the one proposed in [260], except that the former does statical analysis for system call gathering and the later uses dynamic analysis. In [260], the authors proposed a finite state automaton to record system call sequences and a runtime training phase is performed to train the model. However, in [304] the control flow graph of the application is analyzed to generate the automaton. In both approaches, during runtime, the behavior that deviates from the automaton is considered an anomaly. Since the automaton is generated statically in [304], it could contain execution paths that in reality could never be executed. On the other hand, static analysis will detect all program paths, even the ones that a dynamic one might miss if they are not followed during the analysis stage. The third approach in [304] uses non-deterministic pushdown automaton to model the system call sequences with the extension of eliminating all the program paths, those will never be executed. The fourth approach proposed in [304] uses techniques similar to that of [93], aside from the fact that [304] uses static techniques to build the model while [93] uses dynamic techniques. In [93], the authors proposed the use of a sliding window to record system calls during a training period, when the sequence of system calls of a specific size is recorded. During the actual runtime, the same window is slided along the system call trace of the application and a mismatch there is flagged as an anomaly.

In [307], Warrender et al. use sequences of system calls into the kernel of an operating system, one of a wide variety of observable data to distinguish between legitimate and illegitimate activities, for intrusion detection in systems. The authors have used different data modeling approaches to summarize the normal behaviour of a system accurately, based on system traces from different applications, and therefore are able to detect intrusions. The data modeling approaches used are (1) simple enumeration of observed sequences; (2) comparison of relative frequencies of different sequences; (3) a rule induction technique; and (4) Hidden Markov Models (HMMs). They conclude that a weaker method than HMMs is a sufficient solution to this problem.

In [130], Joglekar et al. have proposed an anomaly based Intrusion Detection System (IDS) that could be used to detect malicious use of cryptography and other such applications. The authors claim that their system is unique in aspects such as the ability to monitor cryptographic protocols, and application-level protocols embedded in encrypted sessions as a lightweight monitoring process, and reacts to protocol misuses through directly modifying the manner in which the protocol behaves and responds [130].

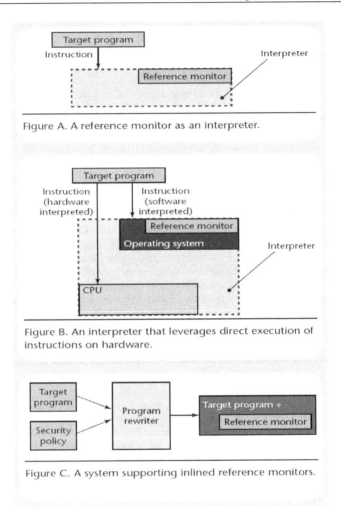

Figure A. A reference monitor as an interpreter.

Figure B. An interpreter that leverages direct execution of instructions on hardware.

Figure C. A system supporting inlined reference monitors.

Figure 3.21: *Reference Monitors for Policy Enforcement (taken from [252])*

3.3.6   Sandboxing

The Principle of Least Privilege was originally formulated by Saltzer and Schroeder in [247] as: *"Every program and every user of the system should operate using the least set of privileges necessary to complete the job"*. The idea behind this principle is to allow just the minimum possible privileges to authorize a legitimate action, in order to enhance protection of data and functionality from malicious behaviour [252]. Sandboxing is a popular method for developing confined execution environments based on the principle of least privilege, which could be be used for running untrusted programs. A sandbox limits or reduces the level of access its applications has on the system. Sandboxes have been of interest to systems researchers for a long time. Butler Lampson, in his 1971 paper [149], proposed a conceptual model highlighting properties of several existing protection and access-control enforcement mechanisms. Even though, Lampson used the word "domain" to represent protection environment, his semantics were equivalent to sandboxing.

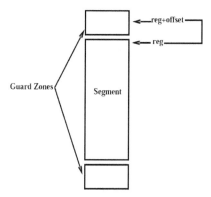

Figure 3.22: *A Segment with Guard Zones (taken from [306])*

Generally, sandboxing in computer systems is achieved through either fault isolation or policy enforcement. Fault isolation makes certain that a failure in part of a software does not cause a complete failure to the system. The most commonly used fault isolation model is the loading of the code into isolated hardware enforced address space. However, the execution overhead for this naive approach is considerably high due to expensive context switching caused by the address space isolation. Policy enforcement ensures safety by dictating what a specific application can and cannot do as policies. Generally this is done via reference monitors as depicted in Figure 3.21.

The first known concept of Software base Fault Isolation (SFI) was proposed in [306]. SFI uses techniques which are different to that of separate address spaces and therefore reduces context switching overheads. In SFI all the jump or write instructions that access an address is instrumented with a check that will verify whether the address is within the guard zones as depicted in Figure 3.22. The size of the guard zone covers the range of possible immediate offsets in register-plus-offset addressing modes. Even though this technique reduces the vulnerabilities and makes an attack harder to the attacker, it will probably not be able to stop, for example, a buffer overflow that overwrites the return address of a function at the next locations of the memory.

In [271], Small et al. extends SFI to formulate an extension called MiSFIT. MiSFIT performs sandboxing on instructions such that they are allowed to perform read and write, only on the memory regions to which they have access. Figure 3.23 depicts an example transformation applied in the implementation of MiSFIT for loads and stores that use an indirect address. In this example, the region tag is the top sixteen bits of the address and has the value *'0xabcd'*. In the first example, the original address is invalid, so the fault-

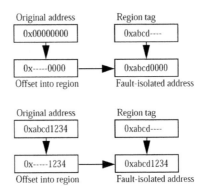

Figure 3.23: *An Example for Transformations performed in MiSFIT (taken from [271])*

isolated address is different. In the second example, the original address is within the
region, so the fault-isolated address is same as the original address.

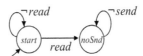

Figure 3.24: *An automaton for 'sent no messages after reading a file' (taken from [80])*

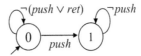

Figure 3.25: *An automaton for 'push once and only once, before returning' (taken from [80])*

Security Automata SFI Implementation (SASI) is a policy enforcement based sandboxing technique which generalizes SFI to any security policy that is defined by a security
automaton. Figures 3.24 and 3.25 depict two examples of such automatons taken from
[80]. Merging-in a security automaton specification (given in Figure 3.25 into an instruction routine is explained in Figure 3.26. The merging involves four phases as depicted in
Figure 3.25 and they are: (1) insertion of security Automata before each target instruction; (2) evaluation of transitions based on the instruction that follows (3) simplifying the
Automata by deleting any irrelevant translations; and (4) compiling the Automata.

Figure 3.27 depicts the security automaton for memory protection, which is implemented
as a prototype for an x86 architecture using SASI. It is interesting to note that the memory protection policy given by the security Automata of Figure 3.27 is equivalent to that
implemented in MiSFIT [271].

In [253], Schneider et al. propose an execution monitor to examine the system at runtime
to see whether certain predefined security policies are enforced. The proposed solution

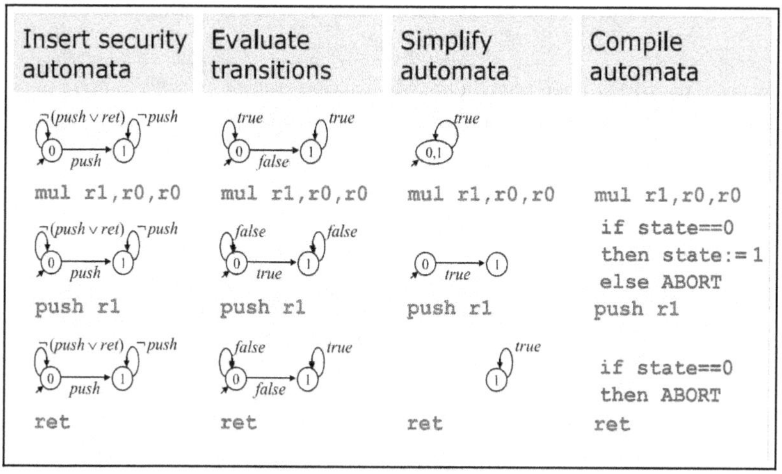

Insert security automata	Evaluate transitions	Simplify automata	Compile automata

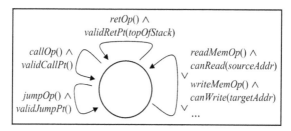

Figure 3.26: *Merging-in a security automaton into a code routine in SASI (taken from [80])*

Figure 3.27: *A security automaton for SFI-like memory protection in SASI (taken from [80])*

does neither need compile time instrumentation nor static source code analysis. These security reference monitors could be implemented at different levels, such as the kernel, firewall, etc.. Schneider et al. also points out that the execution monitors as proposed in the paper can only enforce security policies that are safety features. He also defined security Automata which are similar to that of SASI.

In [84], Evans et al. introduce a system architecture, Naccio which could be used to define safety policies for applications. The policy enforcement is achieved in [84] by implementing envelopes around relevant system calls and by binding the enveloped versions at load- or run-time. Naccio will not stop code injection attack, because only the enveloped system calls are checked at run-time and it is possible to perform code injection attack without using the system calls. In Figure 3.28, an overview of the Naccio system architecture is presented. The top half of the figure depicts what a policy author does to generate a new policy and the bottom half shows what happens, when the first time a user elects to execute a given program enforcing that policy.

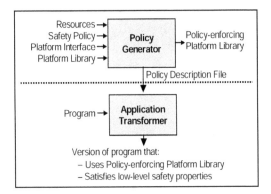

Figure 3.28: *Naccio Architecture (taken from [84])*

In [103], Goldberg et al. propose a policy enforcing monitor, called Janus * that will monitor untrusted applications and disallows system calls that the application is not permitted to execute. The decision as to whether or not to allow a specific system call is made based on the arguments to the system call. For example, a call to open could be a harmless request to open a file the application is allowed to access, or could be a request to open a sensitive file on the hard disk. Janus framework allows a developer to implement and enforce security policies for applications at deployment, that can prevent an attacker from executing system calls, that the application would not normally need to execute in regular operation.

In [218], Provos et al. propose a policy enforcement technique called Systrace for UNIX-based operating systems. Systrace introduces a way of eliminating the need to give a program full administrator privileges (as needed by some system calls), instead allowing finer grained privileges to be given to an application: which system calls can be executed and with which arguments, with much the same functionality as offered by Janus. Overview of system call interception and policy decision is depicted in Figure 3.29. For an application executing in the sandbox, the system call gateway requests a policy decision from Systrace for every system call. The in-kernel policy provides a fast path to permit or deny system calls without checking their arguments. For more complex policy decisions, the kernel consults a user space policy daemon. If the policy daemon cannot find a matching policy statement, it has the option to request a refined policy from the user. Defining a security policy for a program on its potential system call executions is hard and complex, therefore Systrace allows a training algorithm, allowing a user to perform training runs before the program is actually ready to use. Another policy enforcing tool for UNIX-like operating system, FMAC is proposed in [217]. FMAC implements a file system to sandbox applications and it runs in passive (gathers information on the files that the application opens and adds these files to an access list) and active (only file accesses to files on the list are allowed. A file access to a different file will return a non-existent file error) modes.

---

*The name of the Roman God of entrances and exits

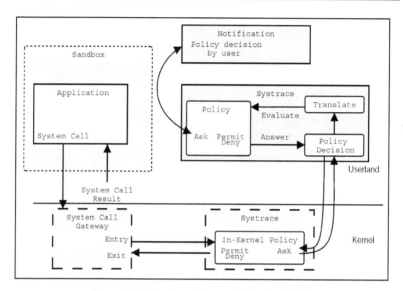

Figure 3.29: *Overview of Systrace (taken from [218])*

In [142], Kiriansky et al. propose a policy enforcement monitoring technique called program shepherding which will monitor program execution and reject control flows that are
considered unsafe. Program shepherding provides three techniques as building blocks for
security policies. Shepherding (1) can restrict execution privileges on the basis of code
origins; (2) can restrict control transfers based on instruction class, source, and target; and
(3) guarantees that sandboxing checks placed around any type of program operation will
never be bypassed.

Figure 3.30 summarizes the contribution of each shepherding technique towards stopping
security attacks. The three boxes represent the three components. A filled-in box indicates that the component can completely stop the attack type above. Stripes indicate that
the attack can be stopped only in some cases. The vertical order of the techniques indicates the preferred order for stopping attacks. If a higher box completely stops an attack,
we do not invoke techniques below it. For example, sandboxing is capable of stopping
some attacks of every type, but we only use it when other techniques do not provide full
protection.

3.3.7   Compiler support

Compilers play a vital role in making the programs written based on language specifications to run on hardware. Compiler is the most convenient place to insert a variety of
solutions and countermeasures without changing the languages in which vulnerable programs are written. The observation that most of the security exploits are buffer overflows
and are caused by stack based buffers, has made researchers propose stack-frame pro-

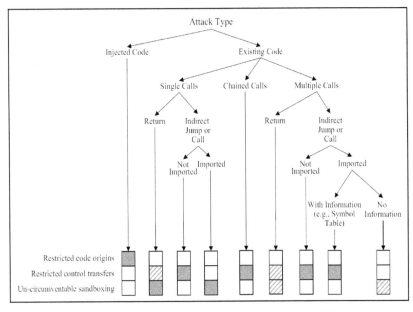

Figure 3.30: *Capabilities of program shepherding (taken from [142])*

tection mechanisms. Protection of stack-frames is a countermeasure against stack based buffer overflow attacks, where often the return address in the stack-frame is protected and some mechanisms are proposed to protect other useful information such as frame pointers. Another commonly proposed countermeasure is to protect program pointers in the code. This is a countermeasure which is motivated by the fact that all code injection attacks need code pointers to be changed to point to the injected code. Since buffer overflows are caused by writing data which is over the capacity of the buffers, it is possible to check the boundaries of the buffers when the data is written to prevent buffer overflow attacks. Solutions proposed as compiler support for bounds checking are also discussed in this section.

Stack Shield, a stack-frame protection technique which is a development tool that adds protection from stack smashing attacks is proposed in [299]. Stack shield protects against stack based buffer overflow attacks, which target return address of the function. The protection mechanism copies the return address of the function to the DATA segment using function prologs and verifies the same, using function epilogues before returning from the function. If the value copied and the value in the stack are different, the stack shield will terminate the program. Even though stack shield seems to be a straight forward solution to the problem, it has been demonstrated [36, 235] that it is possible to bypass stack shield.

Return Address Defender (RAD) as proposed in in [49], is a technique which has the same basics as stack shield in storing the return addresses in a different location. This location

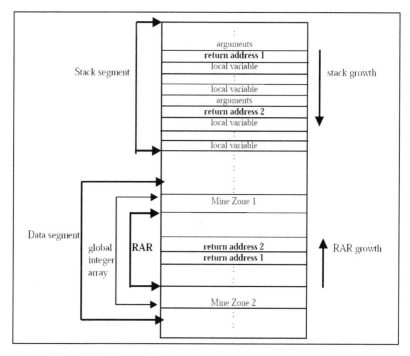

Figure 3.31: *Return Address Repository and Mine Zones (taken from [49])*

is called Return Address Repository (RAR) in RAD. The authors have also proposed mechanisms to protect RAR from being modified. The RAR protection is achieved either via marking the RAR as read only and changing the permission at prologs (where the return address is written to the RAR) or via defining the RAR as an integer array and making the array boundaries (known as mine zones) read only. Figure 3.31 depicts how these RARs are placed under the data segment of the program using mine zones.

StackGuard [59] is another stack-frame protection technique which is applied as a compiler patch to the GNU C Compiler (GCC). An overview of a solution which is proposed in [59] is depicted in Figure 3.32. The return address in the stack is protected by placing a canary word next to the return address and the word is checked and verified whether it is intact before executing the return instruction. An extension to the *StackGuard* is proposed in [57], where the placement of the canary word is moved next to the previous frame pointer in the stack, which is exactly the location where an overflow starts to destroy the stack. A protection mechanism for function pointers is also discussed in [57]. In [235] the author describes techniques, such as indirect pointer overwriting to bypass *StackGuard*.

In [33, 109], the authors have described a mechanism to protect buffer overflow attacks as implemented in Microsoft Visual Studio .NET framework. Similar to *StackGuard*, this

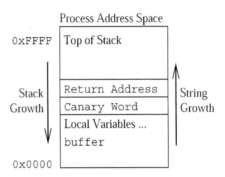

Figure 3.32: *Canary Word next to Return Address (taken from [59])*

technique places a random security cookie next to the previous frame pointer to protect both the saved frame pointer and the return address of a function. Even though, this technique will prevent the normal buffer overflow attacks, a generic bypass (pointer overwriting) to defeat this protection is described in [160].

Another stack-frame protection technique similar to *StackGuard*, known as *Propolice* is described in [82]. *Propolice* tries to prevent the bypasses for *StackGuard* like solutions described in [36, 143, 235]. The extensions are:

1. local variables are re-ordered such that the buffers are placed after pointers to prevent corruption of pointers; and

2. pointers in function arguments are copied to an area preceding local variable buffers to prevent the corruption of pointers that could be used to corrupt arbitrary memory locations.

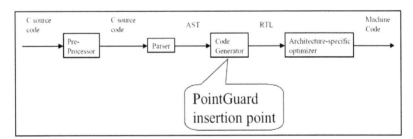

Figure 3.33: *The placement of* PointGuard *in the context of a compiler tool-set (taken from [58])*

In [58], Cowan et al. have proposed a technique to countermeasure pointer overwriting. The technique (which is known as *PointGuard*) attempts to protect pointers by encrypting them while they are in memory and decrypting them when they are loaded into registers

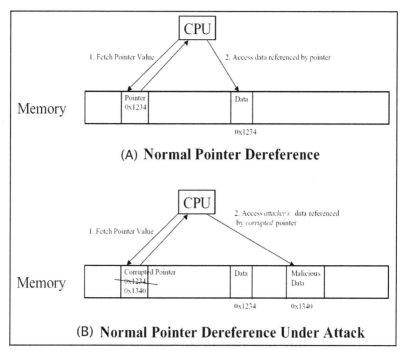

Figure 3.34: *Pointer overwriting when* PointGuard *is not present in the system (taken from [58])*

where they are safe from overwriting. Figure 3.33 depicts the insertion point of *Point-Guard* into a compiler tool-set. Figure 3.34 depicts pointer overwriting when PointGuard is not present in a system. An attacker could change a pointer to dereference an address where the malicious code is injected. Figure 3.35 depicts a system's behavior when *Point-Guard* is present. Even though, an attacker will still be able to modify pointers stored in memory, when the pointers that the attacker provides are decrypted, they will point to different, potentially inaccessible memory locations.

In [325], a security enforcement tool that adds runtime protection against invalid pointer dereferences is proposed. The tool uses static analysis to identify potentially dangerous pointer dereferences and memory locations that are legitimate targets of these pointers. It is implemented as a source-to-source translation that adds run-time checks to pointers after determining, through static analysis, that they are possibly unsafe. If at runtime the target of an unsafe dereference is not in the legitimate set, a potential security violation is reported, and the program is terminated.

In [14], Austin et al. have proposed a pointer and array access checking technique that performs simple program transformation to provide desired error coverage. Figure 3.36 shows the experimental framework of the proposed technique. C programs are translated to their safer counterparts by rewriting all pointers and arrays. When an attempt is made

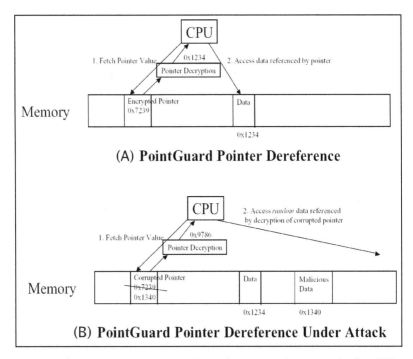

(A) PointGuard Pointer Dereference

(B) PointGuard Pointer Dereference Under Attack

Figure 3.35: *Pointer overwriting when* PointGuard *is present in the system (taken from [58])*

to dereference a safe pointer, two separate verifications are performed and they are: (1) check if this pointer is still a valid one; and (2) make sure that this pointer does not access memory past its bounds. A prototype implementation of this technique is presented in [14]. The overheads of this prototype implementation are measured with six, pointer intensive applications, their overheads are 130% to 540% performance penalty, and around 100% code size increase. As the technique proposed in [14] is not backwards compatible[†], a solution similar to that of [14] with backwards compatibility is proposed in [131].

In a recent paper [72], Dhurjati et al. claim that the problem of enforcing the right usage of array and pointer references in C and C++ programs still remains unsolved. They claim that the approach proposed in [131] and later extended in [245] has extremely high overheads and therefore not worth implementing. In [72], the authors propose a set of techniques that dramatically reduce the overhead of the approach originally proposed in [131], by exploiting a fine-grain partitioning of memory called Automatic Pool Allocation. They show that the memory partitioning is key in bringing down the overheads and their technique successfully detects all buffer overflow violations in a benchmark run they have performed.

In [279], Steffen has added run-time array subscripts and pointer bounds checking to

---

[†]backwards compatibility is the ability to mix checked code and unchecked libraries [14]

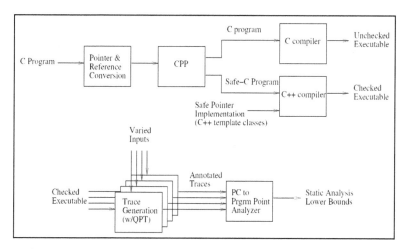

Figure 3.36: *Experimental framework of the technique proposed in [14] (taken from [14])*

the Portable C Compiler (PCC) to form a system called RTCC (RunTime Checking Compiler). Memory pointer overwriting vulnerabilities are captured as they are written instead of when the overwritten memory is used later in the program execution. In the implementation, pointers are represented by three times their normal value, containing the current value of the pointer and the memory addresses of its lower and upper bounds. The RTCC is used to find the true cause of a core dump and to prevent runtime errors. The overheads due to RTCC are around 40% increase in compile time, 3x memory penalty and 10x performance penalty.

In [158], Lhee et al. have presented a technique that performs range checking the referenced buffers at runtime to prevent buffer overflow attacks. Executable files are instrumented with type information of buffers through a compiler extension. As depicted in Figure 3.37, the type information of buffers are read by precompiling the source file with debugging option turned on, and parsing the resulting debugging statements. A type table is generated as a data structure that associates the address of each function with the information of the function buffers. Each object file is given a contractor function so that it associates its type table with a global symbol as shown in Figure 3.37.

In [207], Oiwa et al. have proposed a memory safe ANSI C implementation, known as Fail-Safe C to detect and prevent all unsafe memory operations and to conform to the full ANSI C standards. The implementation consists of sophisticated pointer representations that contain type and size information at runtime. Every memory access operation in Fail-Safe C must ensure that the offset and the type of a pointer are valid. To check this property at runtime, every contiguous memory region also contains its size and type information [‡]. The runtime of Fail-Safe C keeps track of them by using custom memory management routines.

[‡]As depicted in Figure 3.38, each memory block consists of a block header and a data block. A block header contains information on its size and its dynamic type, which we call data representation type

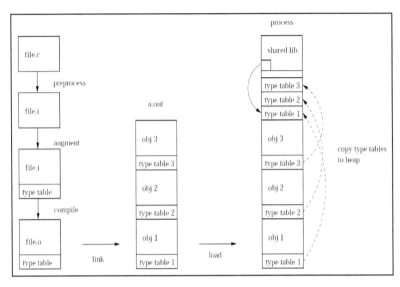

Figure 3.37: *Compilation process, the executable file and the process (taken from [158])*

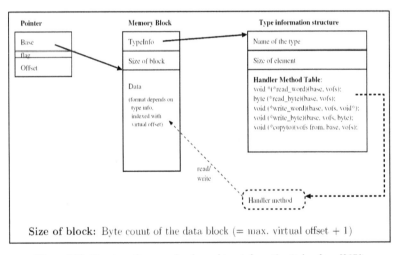

Figure 3.38: *Structure of memory header and type information (taken from [207])*

3.3.8   Library support

Most of the buffer overflows are caused by the mishandling of the standard C library functions, which manipulate strings. Therefore, the obviously naive solution to this problem is to design safer library functions. Safe library functions attempt to prevent vulnerabili-

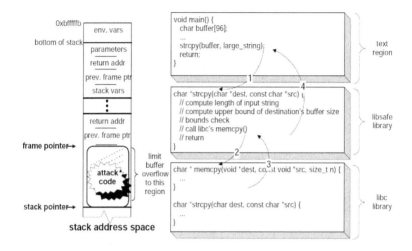

Figure 3.39: Libsafe *Containment of Buffer Overflow (taken from [20])*

ties by proposing new string manipulation functions which are less, or not vulnerable to exploitations. In [186] the authors propose alternative string handling functions to the existing functions, which assume strings are always NULL terminated. The new proposed functions also accept a size parameter apart from the strings themselves. In [174], another safe string library, *SafeStr* is proposed as a replacement to the existing string library functions in C which are immune to buffer overflows and format string vulnerabilities. The goal of the *SafeStr* library is to provide a rich string-handling library for C that has safe semantics. The security goals of the library are to: (1) make buffer overflows impossible; (2) disallow format string exploitations; and (3) track whether strings are 'trusted'.

Another type of countermeasure using libraries is to perform reasonableness checks within the vulnerable string manipulation functions when they are executed. This is achieved by modifying the vulnerable functions in the libraries. A format string based countermeasure called *FormatGuard* has been proposed in [56]. *FormatGuard* is proposed as a patch to the *glibc* library to protect *printf* based format string vulnerabilities. The authors show that *FormatGuard* is effective in protecting several real programs with format vulnerabilities against live exploits. In [20], a method similar to *FormatGuard* is proposed, but to intercept library functions that lead to stack based buffer overflows and are verified before execution.

Two transparent runtime defences against stack based buffer overflow attacks, *libsafe* and *libverify* have been proposed in [20]. *Libsafe* intercepts all calls to library functions which are known as vulnerable, and substitute these calls with alternative functions with the same functionality but that enable buffer overflows to be constrained within the current stack frames. Figure 3.39 depicts a process that has been connected with *libsafe*. The figure shows how the new *strcpy* implementation under *libsafe* is operated by checking for buffer boundary before calling the *memcpy* library function of standard C library.

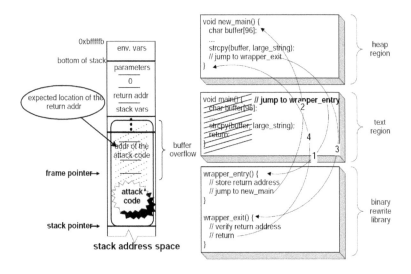

Figure 3.40: *Memory Usage for* libverify *(taken from [20])*

On the other hand, *libverify* modifies the binaries of processes in the memory to ensure that the critical stack elements are verified before they are used. Figure 3.40 depicts the memory of process which is connected to *libverify*. Before the user code is executed, it is linked to the *libverify* library and the binary of the user code is rewritten such that when critical stack elements such as the return address are used, it will be verified by the *libverify* code. There exists a patch [272] to the FreeBSD implementation of the *libc* that performs integrity checks before executing several vulnerable library functions.

In [240], Robertson et al. propose a countermeasure for protecting dynamically allocated memory that attempts to protect against attacks on the *dlmalloc* library management information which is explained in Section 2. Allocated and unallocated memory segment layouts are changed by including a padding mechanism and a checksum into memory management information as depicted by Figure 3.41. The checksum is inserted when a heap-segment is allocated and is verified when it is freed. Therefore, an attacker overwriting the management information with a heap based buffer overflow is prevented. The solution in [240] has been proposed as a patch to the *libc* to detect heap-based overflows at runtime. *ContraPolice*, a technique to protect heap memory regions by placing randomly generated canaries is proposed in [148].

HEALERS, first introduced as a boundary checking mechanism to prevent heap based buffer overflows in [86], later was improved into fault-containment wrappers which makes an application secure and robust [87, 88, 89]. The major function of these wrappers is to intercept calls to C libraries and replace them with versions that do checking on arguments to ensure correct use of the API. Figure 3.42 depicts the generation process of one of these wrappers, the robustness wrapper, which sits between an application and its shared libraries. The wrapper generation, in general is described in Figure 3.43, utilizes two tools

Figure 3.41: *Modified heap-memory segment structure and layout (taken from [240])*

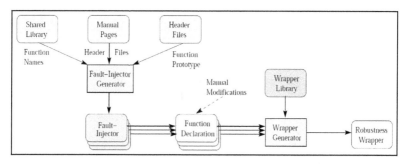

Figure 3.42: *Architecture of the robustness wrapper generation process (taken from [87])*

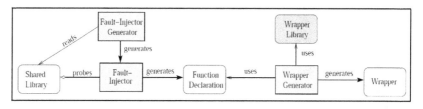

Figure 3.43: *Architecture of the wrapper generation process (taken from [88])*

and they are: (1) fault injection generator; and (2) wrapper generator. Figure 3.44 shows different wrappers such as the security wrapper, the robustness wrapper, etc. As depicted in Figure 3.44, different applications may use different wrappers and can share them.

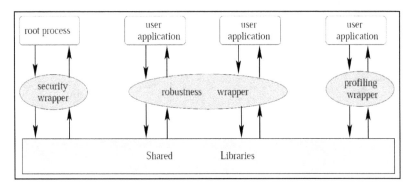

Figure 3.44: *Different Wrappers in HEALERS (taken from [89])*

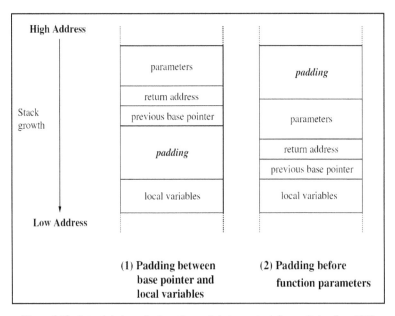

Figure 3.45: *Potential places for inserting pads between stack frames (taken from [27])*

Address randomization technique is proposed in different levels (at the hardware architecture, the kernel and at the library). In [27] Bhatkar et al. have discussed address concealment techniques to make it harder for an attacker to exploit vulnerabilities. The authors have considered many different memory randomization techniques such as stack base address randomization, DLL base address randomization, code/data segment randomization, random stack frame padding and heap randomization. The potential locations for random stack frame padding is depicted in Figure 3.45.

3.3.9   Operating System based Countermeasures

Similar to architecture based countermeasures to code injection attacks, operating system based solutions, use the observation that most attackers wish to execute their own code and have proposed solutions preventing the execution of such injected code. Most of the existing operating systems split the process memory into at least two segments, code and data. Marking the code segment read only and the data segment non executable will make it harder for an attacker to inject code into a running application and execute it.

In [70], Designer has proposed a non-executable stack to prevent buffer overflow that injects code into a stack based buffer. The author has used the Linux kernel on the IA32 architecture to implement the non-executable stack. This approach has the problem of potentially breaking legitimate uses of stack. For example (1) functional programming languages generate code during run-time and execute it on the stack, (2) GCC uses executable stacks as function trampolines for nested functions, and (3) Linux uses executable user stacks for signal handling. Even though it is possible to detect legitimate uses of stack and dynamically re-enable execution, it will open a window of vulnerability and is hard to do it in a general way. Furthermore, while setting the stack non-executable will not cost anything, it could be easily bypassed as explained in [318]. In [210], the authors propose a kernel patch for Linux that implements a non-executable stack and heap by using non executable pages for the IA32 architecture. The NetBSD, starting from version 2.0 supports non-executable mappings [291] on platforms where the hardware allows it.

Most of the naive exploits expect the memory segments to start at a specific location all the time and use this information to overwrite interested addresses such as return address of a function. Therefore, making the base or starting address of a process, random for each program execution, making it harder for an attacker to exploit an execution flow and inject code for execution. In [321, 322] the authors propose Transparent Runtime Randomization (TRR), a modified dynamic program loader, which is used to load programs into memory for execution. Figure 3.46 summarizes the operations performed by TRR when an application (*netscape* as an example) is launched. In [48], the authors have proposed another operating system based randomization technique that randomizes system call mappings, making it harder for an attacker to call system calls in injected codes.

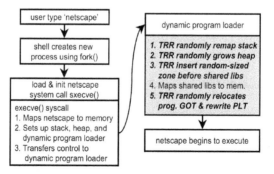

Figure 3.46: *Operations of TRR at a Process Launch (taken from [322])*

As most of the buffer overflows try to overwrite the return address of a function, there are many architecture based countermeasures to protect return addresses as described under architecture based countermeasures. In [96], an operating system based countermeasure, called *StackGhost*, to protect return addresses is proposed. *StackGhost* is a patch to the OpenBSD kernel for the SPARC architecture. *StackGhost* transparently and automatically protects the return addresses of the applications. The protection mechanisms require no application source or binary modification and imposes only a negligible performance penalty according to the authors of [96]. Interested readers are referred to [311, 312] for a further detailed analysis and comparison of these countermeasures.

3.4 Reliability Support in Computing Systems

Most of the known reliability support systems rely on the control flow of the application that runs on them to detect non-reliable behaviours due to either transient faults or permanent errors. The right control flow is a primitive part of the correct execution of computer programs and it is also a necessary condition to all the software implemented error detection techniques. Therefore, Control Flow Checking (CFC) is the prominent checking mechanism when it comes to reliability support. For more than three decades, different control flow checking mechanisms have been used to verify proper flow of a program. Few of the first known publications on control flow error detection include [225, 324], where the authors outline a general software assisted scheme for control flow error detection. However there are a few proposals to use other mechanisms such as memory access checking [195], checking of control signals [63, 124, 277] and reasonableness checking [167, 169].

In this book the literature on reliability support provided for computing systems is divided into three categories based on the type of support the systems provide and they are:

1. Control Flow Checking using Watchdog Processors;

2. Control Flow Checking with other hardware based techniques;

3. Software based Control Flow Checking Schemes; and

4. Soft Error Detection and Recovery.

3.4.1 Watchdog Processor based Control Flow Checking Schemes

Watchdog processors are comparably smaller co-processors, which are used to perform simultaneous error detection by observing the aspects of main processors [162, 168]. Watchdog processor is the cost effective solution for on-line error detection at system level, in general purpose processor systems built upon commercial off-the-shelf (COTS) hardware components. Historically, the idea of using watchdog processors are extended from watchdog timers [54, 208], which were used to monitor the timing of system activities concurrently with the activities themselves.

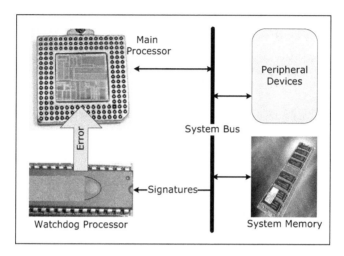

Figure 3.47: *Watchdog Processor*

Figure 3.47 depicts an overview of how a watchdog processor interacts with the main processor. The watchdog is provided with information about the states of the processor at the start and watchdog at runtime checks the main processor's behavior to detect any problems. When an error is detected, a signal is either sent to the main processor or a device that handles the error recovery or remedy. The following are the reasons given by watchdog advocates for using it.

1. Concurrent error checking avoids significant performance penalties.

2. Less hardware redundancy compared to replication of the main processor.

3. Design diversity[§] is good to avoid common mode errors [15, 329].

4. Can be added to existing processors designed with COTS components.

5. Could be used not only to detect transient hardware errors, but also to identify and recover from errors caused due to software or design errors.

6. Could be used as a recovery system by storing error-specific and application specific information.

Even though the watchdog processor advocates provide the above benefits, the design of such system is restrained by the following factors:

1. The checking mechanism has to be performed concurrently with the main processing, and otherwise it will result in performance degradation.

[§]Watchdog processors are designed by different manufacturers than those who design the main processors

2. The interface between the watchdog and the main processors has to be interfaced easily without major modifications.

3. The watchdog processor has to be very simple compared to the main processor, so that the reliability of the watchdog processor could be trusted more, compared to the main processor.

Watchdog process based control flow checking schemes could further be divided into two groups based on the representation of the reference information and in the derivation of the runtime information, and they are: (1) derived runtime signature base control flow checking [3], and (2) assigned runtime signature based control flow checking. They are discussed in detail below.

Derived Signature Based CFC

Under the derived signature based control flow checking technique the signatures are computed for logical basic blocks of an application, based on instruction sequences through information compaction such as check-summing. At runtime, similar signatures are calculated by the watchdog processor concurrently and they are compared against reference signatures given earlier. The first known technique of this kind is proposed in [194], where a control flow graph is divided into a sequence of branch free nodes, called *a path set* which are then assigned with a unique signature and compared against the runtime signature. This approach, which is called Basic Path Signature Analysis (BPSA) needs a complicated parser to generate the path set and then the signature.

Over the years, a considerable number of improvements and changes have been suggested to the basic idea presented in [194]. These changes consider improving one or more of the design factors such as the complexity of checking [277, 298] , design and runtime overheads [314, 315], error detection latency [205, 249, 274], etc.

In [165] a hardware watchdog scheme with a signature monitoring technique called On-line Signature Learning and Checking (OSLC) is proposed. Unlike traditional watchdog based derived signature based CFC, the signature generation is performed during a runtime learning phase in [165]. The benefits of this idea are: (1) reference signatures are calculated at runtime and the information they represent are towards the real execution of the system, and (2) compiler/assemble support is not necessary to perform these checks. However, the authors have only tested this system with very simple applications and a non-complex hardware architecture. An on-line control flow checking technique for a controller with a self-testing structure is discussed in [81].

Assigned Signature Based CFC

Under assigned signature based CFC, the signatures for each node in a control flow graph of an application are assigned arbitrarily such as using prime numbers and loaded into the watchdog processors using signature transfer instructions. The benefits of this scheme are: that it is an easy implementation of the software instrumentation process of such a system; and it is possible to perform runtime checking asynchronously. However, there are a couple of major drawbacks on using this scheme and they are: (1) performance degradation due to explicit signature transfers from main processor core to the watchdog processor at runtime, and (2) very low error coverage, since only the sequence of the code is checked and not the content of the application code. The first known proposal of assigned signature based CFC is published in [324], where specific points in the execution sequence were assigned with distinct prime numbers, and checked at runtime for verification.

According to Majzik et al. in [170], the first commonly known typical software configuration method that was used in all assigned signatures method, known as Structural Integrity Checking (SIC) was introduced in [162]. Extended Structural Integrity Checking [179, 180] extended the checking capability of SIC, and included checking mechanisms for runtime computed procedure calls and interrupt handlers.

Some researchers have used a different approach to classify watchdog processor based control flow checking schemes. They categorize the systems depending upon how the static signatures are stored and accessed. The first category embeds signatures into the application binary itself [67, 68, 162, 254, 255, 313] and the second category uses a separate memory (dedicated memory of the watchdog processor) to store and access the signatures [180, 181]. In [255] the signature is embedded into the application instruction stream at the assembly level, and branch address hashing is used to reduce memory overhead. In [180, 181], the signature calculated at compile time is stored in a separate memory belonging to a watchdog processor and therefore the original application/program does not have to be modified. Another category of watchdog processor based systems which was not discussed above, is the watchdog processors used in parallel or multiple processor systems. Examples of such systems are described in [79, 179, 223].

3.4.2 Other Hardware Based Control Flow Checking Schemes

In this section, the author discusses hardware assisted control flow checking mechanisms which do not use watchdog processors to perform the checking. The advocates for the techniques of this kind argue that they are able to overcome the drawbacks of watchdog processors which are discussed in section 3.4.1.

A concurrent, on-chip hardware assisted control flow checking technique is proposed by Leveugle et al. in [157]. The technique in [157] uses control signatures to enable control flow checking. However, Leveugle et al. prohibit indirect register branches as they are unpredictable at compile time. Therefore Leveugle et al.'s technique is not general enough to be applied to most of the embedded processors. Another hardware assisted concurrent error-detection method is proposed in [97] for embedded space flight applications. Authors in [97] use parity checks in registers and signatures for CFC (*XOR*ing instructions until a check-point, where they are verified). Multiple bit errors (or bursts) are not captured by this technique and signatures will not reveal the exact point of the error in the flow of an application.

A multiple input signature register (MISR) based control flow checking mechanism is

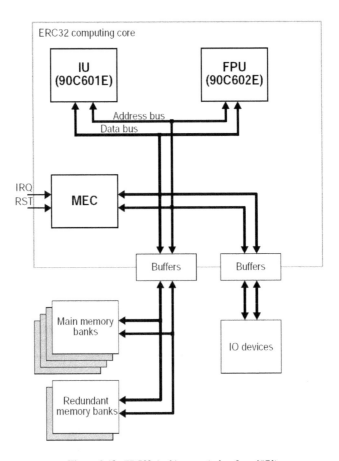

Figure 3.48: *ERC32 Architecture (taken from [97])*

proposed in [316]. Figure 3.49 depicts the signature placement of this technique. The left side of the figure shows the basic technique where the compiler places a reference signature instruction at the end of each basic block and during program execution, a simple hardware monitor regenerates the signature and checks it at each reference signature. The right side of the figure shows the justifying signature technique where the compiler sets the signatures so that the signatures of two paths are consistent at the location where the paths merge. Figure 3.50 depicts the datapath for a RISC processor and the datapath for the corresponding signature monitor as detailed in [316]. As stated earlier MISR is used for runtime signature calculation.

Even though various hardware based monitoring techniques have been proposed in the literature for the purpose of on-line testing in general and control flow checking in particular, it is unfortunate that they are seldom implemented in real commercial microproces-

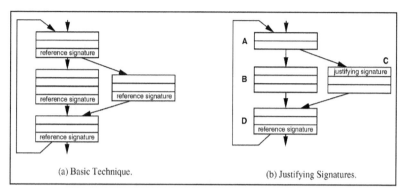

Figure 3.49: *Placement of Program Signatures (taken from [316])*

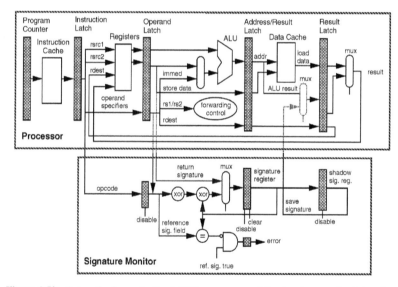

Figure 3.50: *Datapaths for a pipelined RISC processor and its signature monitor(taken from [316])*

sors. However some recent processors, such as Intel's Pentium processor, have provided means for monitoring their hardware through clock and other programmable (could be programmed to count particular events in a system) counters. Sosnowski et al. in [275] have discussed such counters. In, another recent study [125], Iyer et al. report some of the advances such as their RSE framework [192] in current processors that support hardware-level reliability support.

3.4.3   Software Based Control Flow Checking Schemes

Software control flow error detection is performed by having appropriate signatures for similar blocks as per hardware techniques, but the checking is done by software code inserted into the instruction vector at compile-time. Since software assertion performs the necessary checking, there is no need for a separate hardware monitor. Software error detection schemes use software routines to check proper control flow at runtime. The routines are inserted into the application at assembly level or at a higher level. Some of the techniques are described in [2, 104, 105, 134, 188, 205].

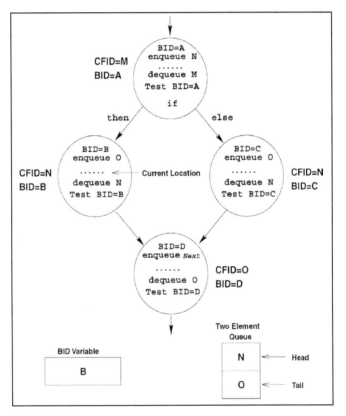

Figure 3.51: *Assignment and checking of IDs for IF-THEN-ELSE construct after the placement of CCA assertions (taken from [2])*

Based on how the checking routines are used, the author has different classifications for software based control flow checking. Control Checking with Assertions (CCA) inserts assertions at the entry and exit points of identified branch-free intervals [134, 176, 191]. CCA is implemented as a pre-processor to a compiler, based on the syntactic structure of the language and does not require any CF graph generation and analysis. Figure 3.51 depicts the procedure of assigning signatures for an IF-THEN-ELSE construct.

An enhanced version of CCA, Enhanced Control Checking with Assertions (ECCA) is proposed in [2] which targets real-time distributed systems for the detection of control flow errors. ECCA addresses the limitations of CCA (higher code size overhead, uncovered CFIs in the assertions, etc.), and is implemented at both high and intermediate levels (register transfer language) of a language.

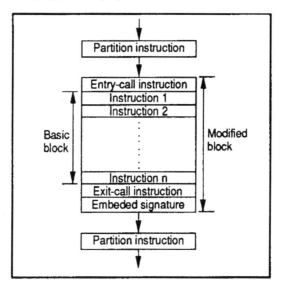

Figure 3.52: *An overview of the BSSC mechanism (taken from [188])*

Another type of software based control flow checking technique, called Block Signature Self Checking (BSSC) is proposed in [188]. An application program is divided into basic blocks and each of these are assigned a signature. A set of instructions (assertions) at the end of a basic block reads the signature from a runtime variable and compares it to an embedded signature following the instructions. A mismatch in the comparison indicates a control flow error. An overview of the BSSC mechanism is depicted in Figure 3.52.

PreEmptive COntrol Signature (PECOS) checking [19] is the only software based *preemptive* control flow error detection technique available to date. PECOS is proposed for use in a wireless telephone network controller to detect control flow errors. Preemptive error detection gives us the guarantee that the error is detected before an erroneous control flow instruction is executed. PECOS uses assertions formed with assembly instructions that can be embedded within the assembly code of an application. PECOS detects control flow errors caused by a direct or an indirect jump instruction. Due to the extensive use of assembly instructions as assertions, PECOS suffers from high memory overheads and reduced performance. Furthermore, since PECOS's software instrumentation is performed at the assembly level, the library functions have to be instrumented separately. Therefore, the error coverage of PECOS is affected by the control flow errors introduced in control flow instructions in the library functions.

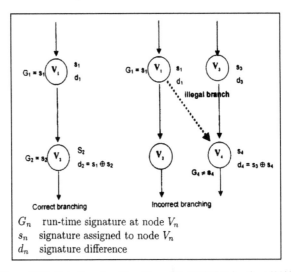

G_n run-time signature at node V_n
s_n signature assigned to node V_n
d_n signature difference

Figure 3.53: *Detection of an illegal branch in CFCSS (taken from [203])*

Another software based control flow detection technique called Control Flow Checking by Software Signatures (CFCSS) is proposed in [203]. CFCSS assigns a signature to each basic block at compile time and uses a dedicated general purpose register to track these signatures at runtime. Figure 3.53 depicts how the CFCSS behaves with a correct branch and an incorrect one. The runtime signature of a node is calculated by XORing the previous node's signature with the signature difference of the current node. A mismatch in the calculated and assigned signature indicates a control flow error. The signature calculation used in CFCSS require all the predecessors of a basic block to have the same signature. This limitation will have implications in this technique's error coverage, such as an error in a branch to a wrong block with the same signature as the right block cannot be detected. In [232, 233] the authors have proposed Enhanced Control-Flow checking (ECF), a technique that extends CFCSS with a runtime adjusting signature so that the above mentioned restrictions of CFCSS are fixed. A software based compile time instruction duplication technique for super-scalar processors, called EDDI is proposed in [204]. EDDI uses different registers and memory variables for duplicating the instructions at compile time.

A dynamic binary translation based control flow error detection scheme is proposed by Borin et al. in [31]. Borin et al. have proposed an error classification mechanism for control flow errors and software based control flow checking solutions to detect the errors. Their solution detects all the errors under their classification. Figure 3.54 depicts the control flow error categories classified by the authors in a control flow graph. The solid lines in the figure are correct control flows and the dashed lines are the branch errors. The proposed software solutions are: (1) the Edge Control Flow Checking (EdgCF) technique, where a special register is loaded with the signature of the next basic block and verified

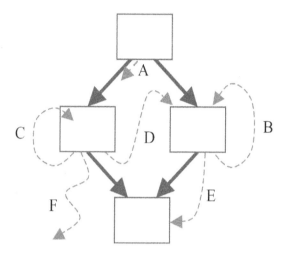

Figure 3.54: *Branch Error Categories (taken from [31])*

at the beginning of that block; and (2) the Region Based Control Flow (RCF) technique, assigns signatures to regions in contrast to basic blocks which are verified similarly.

3.4.4 Soft Error Detection and Recovery

Availability of computer systems depends on hardware and software reliability. Hardware is not able to provide full coverage for faults caused by soft errors, due to the high cost arising from the requirement and scalability issues. Thus the responsibility to handle soft errors is usually left to the system software. But from the literature it is evident that the system software does not have any provision to handle hardware faults raised due to soft errors. Soft error detection and correction are done at different levels starting from hardware level moving towards the application level. Higher levels are capable of detecting the errors which escaped lower levels. The error detection schemes are separated into hardware based methods and software based methods and both are analysed here.

Hardware Support for Soft Error Detection and Recovery

Hardware support for Error coverage has been implemented in the form of complex hardware units and software blocks. Traditionally, this was mainly available only for servers and in few high end commercial processors. The error detection on these servers was enabled by specialised redundancy hardware [22] such as the IBM S/390 Parallel Sysplex [201]. Hardware faults may arise due to the soft errors present in either the memory and communication link or in logic circuits. Research work is in progress to find and correct the errors in logic circuits, at both circuit and architectural levels. The outcomes of such research are at their infant stage, and are too complex to be implemented in commodity processors.

Parity check or Error Correction Codes (ECC) are used for error detection in memory and system buses [159, 239]. ECC are capable of detecting two or more bit errors and correcting single bit errors. The detected errors are corrected by the hardware or firmware. The processor either corrects the error or reports it to firmware, and the firmware requires additional software support to handle the errors. The evolution of error handling feature in memory is slower. The authors in [285] have presented hot plug redundant array of industry-standard dual inline memory modules to achieve high fault-tolerance, increase availability along with increasing memory size. The protection provided by [285] is higher than that of standard ECC methods as it is able to correct double, four and eight bit DRAM errors. Thereby it provides a higher availability, scalability and fault tolerance.

The IA-64 architecture designed for Intel's Itanium processor targeting high end server market provides support for reliability and availability [219]. The support provided in [219] includes fault coverage at the hardware level, a configurable multilevel error containment strategy, an availability feature set and a machine check abort architecture. The memory structures are protected by either ECC or parity. The configurable multilevel error containment strategy prevents error propagation at different levels (system, cluster and process), by using a hardware reset pin at the system level, an error indicator at the cluster level, and a data poisoning mechanism at the process level. Availability features consists of extensive error logging, watch dog timer, and a low priority interrupt. The less critical errors are corrected and the information is communicated with the system firmware and operating system by the machine check abort architecture using firmware and hardware.

There are many proposals to use duplication or redundancy to detect soft errors. A duplicate copy of the program is re-executed either on a special hardware dedicated for this purpose, or on a redundant hardware unit, and the results are compared to check whether there were any faults in the execution [95, 226, 229, 231, 241, 302, 308].

Software support for Soft Error Detection and Recovery

There are several software mechanisms proposed in the literature for software error handling [32, 187, 230]. Each of these is based on the level of the hardware support provided by the processor and the type of the system. Even if the hardware is unable to detect an error, the software scheme will provide means to detect those escaped errors. Some of the software based error detection schemes are discussed below.

Assertions, such as logic statements are inserted into programs in [234] to reflect invariants at runtime and they could be used to detect soft errors. Data and code redundancy [230] could be implemented by modifying the source code with insertion of duplication of copies and checks at proper places. This could be done at the pre-compilation phase. However, the software overhead of this technique is high. Procedure duplication is used in [127] and the results are verified. In [324] basic block consistency is maintained by comparing statically and dynamically generated deterministic signatures.

Fault recovery mechanism varies depending upon the severity of the faults and the level at which they are detected (hardware, software or operating system). The complexity of the recovery software varies depending upon the available hardware support. Based on the detected error, the recovery may be full or partial. When a user signalable error arises, the processor will let the kernel continue to operate, and the kernel will signal the error to the user task. User application will recover according to the availability requirements. For a kernel fatal error, complicated check-points based rollback recovery mechanism could be used, or the program code could be re-fetched.

3.5 Fault Injection Analysis

Fault injection and analysis is used to evaluate fault tolerant designs and to study the fault propagation pattern of a system. The most common fault injection and analysis systems could be divided into: (1) hardware based physical fault injectors; (2) software based fault injectors; and (3) simulation based fault injectors. Another rare method of fault injection analysis is analytical modeling [296]. In analytical modeling state models are built, and are solved with mathematical formulations with (an example is in [246]) or without the help of tools. Even though these analysis are good to model systems, when the system parameters are known, it is hard to evaluate parameters so that modeling could be performed.

Fault injection analysis as a formal system has been discussed for the first time in [150]. Later, in [10] authors propose a pin-level fault injection tool for dependability assessment and they extend and formalise their model in the following year [8, 9]. This formal model was used in [216] to assess *Delta-IV* architecture, which provides fail-silent nodes using hardware interlocks. Around the same period, the authors of [18] proposed fault injection analysis in the design and test phases of processors to identify and remove faults from the design.

Hardware based physical fault injection tools could further be divided into three groups based on the approach they take to inject faults. They are as follows.

1. Use heavy-ion radiation to inject faults. In [113] and [138, 139] the authors use heavy-ion radiation to inject faults into one of many replicated processor boards and evaluate the effect of radiation.

2. Use pin-level fault injectors to inject faults. In [164] the authors have presented a tool called RIFLE to inject pin-level faults.

3. Use electro magnetic disturbances to inject faults. In [137], the authors compared all three groups in an architecture called MARS. In [248] the authors use a laser fault injector to inject upsets into transistors in a VLIW processor.

Hardware based fault injection tools closely resemble faults that one would expect from an aggressive environment such as in space. In hardware approaches, the system operates correctly, while the faults are being injected and it deviates only when the errors are manifested (systems of such kind are called non-intrusive). However, most hardware based physical fault injection and analysis systems are expensive and may cause permanent damage to the system.

During the same period when hardware based fault injection tools were proposed, there were proposals for fault injection tools to inject faults into distributed system software. These software based fault injection tools are called, SoftWare Implemented Fault Injection (SWIFI). The first of such tools is Fault-Injection based Automated Testing environment, FIAT ([23, 258]. FIAT will test for trigger conditions and inject faults at compile time. After the introduction of FIAT, the authors of [135] proposed a Unix system call (ptrace) based SWIFI called FERRARI. FERRARI was implemented in SunOS and the user was able to inject faults into memory and registers without re-compiling the application. Orchestra [64] is another SWIFI developed to inject faults into real-time distributed systems. In Xception [39, 40], the authors used the debugging registers in modern processors to trigger faults in distributed PowerPC systems with very limited level of intrusion. In [297], the authors proposed FTAPE, a SWIFI tool to show the fault propagation in different fault tolerant computer architectures. Loki is a state driven fault injector proposed in [46, 60] to inject faults in distributed systems.

There are many simulation based fault injection tools, which are used to evaluate fault tolerant designs. The first known tool, DEPEND was proposed in [107, 108]. DEPEND is a simulation based tool for modeling systems and system-level faults. Extended models of DEPEND were later used in [133, 236]. ADEPT is a simulator based system [100] used to model complex, real-time distributed systems. Evaluation tool for multiprocessor architecture with fault injection analysis is implemented in REACT [51, 52]. FOCUS [50] is a simulation environment for conducting fault-sensitivity analysis of chip-level designs. The FOCUS environment can be used to evaluate alternative design tactics at an early design stage. In MEFISTO [126], the authors propose the integration of a fault injection methodology within the design process of fault-tolerant systems in a VHDL based simulation model.

The SWIFI systems are cheaper to implement than either hardware or simulation based systems, since SWIFI systems neither require special hardware nor simulators. SWIFI also has a benefit over physical systems as we can specify the faults to be injected and the same fault injection can be repeated. Even though SWIFI tools maybe intrusive as the fault injector might need to execute on the same processor, the intrusion level is often low. Although SWIFI fault injection is limited to the processor components those are affected by software, it can be used to inject faults in hardware components which otherwise would not possible with physical injection (for example, a particular register in a general purpose processor). In [61], the authors found that more than one-third of device level random faults manifested in ways that could be expressed by software based fault injectors. Simulation based fault injection analysis has the advantage of being able to inject any fault model. When the source code of a simulator is available to the user, it is the least expensive, simple and quick method of fault injection analysis. However, simulations do not always exactly model the behavior or a real system and simulators do not consider implementation errors that may exist in a real system. For example, in [280] authors compare a simulated fault injection with SWIFI and a SWIFI fault injector on a real system.

The authors of [280] report that when the system state is not well defined (ambiguous specification) or not defined at all, the simulator behaves very poorly on resembling the real system. However, when the system operates in valid state, the simulation was very accurate.

Few researchers claim to use more than one type of analysis. In [328] the authors claim they use both SWIFI and hardware based analysis. In [116], the authors use both SWIFI and simulation based fault injection analysis. In [116] SWIFI is used for finding the target check-points and simulation for injecting faults. In [136], authors combine hardware and software for fault injection and check-point triggering. The authors of [136] also argue the importance of using both hardware and software fault in a validation experiment. In FlexFi [26], the authors use a separate debug-mode micro-controller to inject faults into another micro-processor in real time.

3.6 Summary of this chapter

In this chapter, the authors have presented most of the known countermeasures for security and reliability problems proposed in the literature. He has also presented a thorough analysis of countermeasures for code injection attacks. In the later part of this chapter, he has given details on how fault injection analysis has been used over the years in fault tolerant designs.

Chapter 4

Monitoring Framework

... 'well, I'll eat it,' said Alice, 'and if it makes me grow larger, I can reach the key; and if it makes me grow smaller, I can creep under the door: so either way I'll get into the garden, and I don't care which happens!'

— Lewis Carroll, *Alice's Adventures in Wonderland*

At a given time, a processor executes only a few instructions and large part of the processor is idle. Utilizing these idling hardware components by sharing them to perform security and reliability monitoring is a wise step to reduce the impact of the monitors on hardware cost. Our framework requires little hardware overhead in comparison to having additional hardware blocks outside the processor. This reduction in overhead is due to maximal sharing of hardware resources of the processor. Using micro-instruction routines within the machine instructions, allows us to share most of the monitoring hardware. Our framework is superior to software techniques as the monitoring routines are formed with micro-instructions and therefore reduces code size and execution time overheads, since they occur in parallel with machine instructions.

4.1 About this Chapter

4.1.1 Objectives

The primary objectives of this chapter are:

1. introducing the monitoring framework that is used in all our proposed hardware based solutions to counter security and reliability problems.

87

4.1.2 Outline

Design of embedded systems is constrained by strict requirements in latency, throughput, power consumption, area limitations, cost effectiveness and time to market pressures. A well-known challenge during an embedded processor design is to obtain the best possible results for a typical target application domain. A custom instruction set to perform special tasks could result in significant improvements for an application domain. In recent years, Application Specific Instruction-Set Processors (ASIPs) have gained popularity not only in the research community but also in the production of embedded systems. The existence of rapid design tools for ASIPs makes it easy to incorporate our security and reliability framework into its design process.

4.2 ASIP as the Design Framework

Embedded systems have become an intrinsic part of many modern devices such as mobile phones, digital cameras, cars, etc. A recent article on embedded processors indicates that the embedded processors' market share is around 98% of all microprocessors [129]. Therefore the demand for such systems is exppnentially growing along with their capabilities and complexities. Programmable processors are a feasible solution to cope with this demand at an affordable cost. However, this leads to having software implementations, which have to meet real-time system requirements and be energy efficient. This limitation has brought the need for specialized embedded processors with application specific hardware components in them. A compromise between fully programmable general purpose processors (which consume more power) and non programmable application specific integrated circuit (which are neither flexible nor reusable) is achieved via Application Specific Instruction-set Processors (ASIPs) [4].

Through the introduction of ASIPs, it has become possible to fill the energy-flexibility gap in the previous processor paradigms. Tools such as *ASIP Meister* [293] from PEAS project [4], *XPRES* from Tensilica Inc., etc. empower the rapid and effective design and generation of ASIP models in a hardware description language. *ASIP meister* is capable of generating single issue, RISC* (reduced instruction set computer) ASIPs. The rapid design process enabled by an automatic processor design tool such as *ASIP meister*, is used as the basis for the implementation of our secure and reliable framework. However, it should be noted that the automatic design tool is used only as an implementation tool and our framework is capable of serving any embedded processor architecture and is not restricted to the design tools used to generate it.

Figure 4.1 depicts the general ASIP design flow when an automatic processor design tool, such as *ASIP Meister* is used. Further, Figure 4.1 also depicts the additional steps followed to include our security and reliability monitoring framework. Based on a RISC processor's instruction set architecture (ISA), the instruction formats and the instructions

---

*RISC is a microprocessor CPU design philosophy that favors a simpler set of instructions that all take about the same amount of time to execute. The most common RISC microprocessors are AVR, PIC, ARM, DEC Alpha, PA-RISC, SPARC, MIPS, and IBM's PowerPC [237]

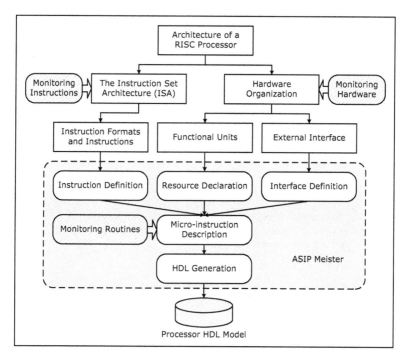

Figure 4.1: *ASIP Design Flow as Our Design Framework*

are decided and given as an input to *ASIP Meister*. The hardware organization of the processor is used to decide on the functional units and the external interface, which are also given as inputs to *ASIP Meister*. Based on the given inputs, *ASIP Meister* defines the instruction, declares the resources to be used and defines the external interface of the processor. Micro-instruction description is used to link the instruction definitions to the resources and the interface. That is the micro-instructions will form the wires and therefore the state changes of a processor for each instruction. Finally, the hardware description language (HDL) generator is used to generate the processor model in HDL.

The ASIP design flow, as described in the previous paragraph is altered as indicated in Figure 4.1 to include the implementation of our security and reliability framework. Based on the type of security/reliability solution, the ISA of the processor is extended with additional instructions[†]. Even though most of the time our monitoring framework uses the existing hardware, few additional hardware components are added for monitoring (if necessary) as shown in Figure 4.1. Including the dominant part of our framework into the ASIP design flow comes as micro-instruction based monitoring routines (depicted in Figure 4.1. The micro-instructions of selected instructions are altered such that they perform security and reliability monitoring in parallel to their regular processing. Further details on specific monitoring routines are explained in the coming chapters, where the security and reliability solutions are explained in detail.

4.3   Summary of this chapter

In this chapter, the authors have presented the monitoring framework which they have developed during the first authors thesis and how this framework uses an ASIP design flow for its implementation.

[†]Instructions are added only when they are needed; for example, in inline security monitoring technique, as described in Chapter 5, no additional instructions are added to the existing ISA

Chapter 5

Hardware Assisted Runtime Monitoring for Inline Security and Reliability Checking

...'Found what?' said the Duck. 'Found it,' the Mouse replied rather crossly: 'of course you know what "it" means.'

— Lewis Carroll, *Alice's Adventures in Wonderland*

Countermeasures for software based security attacks by 'software-only' approaches increase code size, reducing performance and 'hardware assisted' approaches use additional hardware monitors incuring high area penalties. Further, many of the hardware assisted techniques require compiler support to instrument the application programs. The project described in this chapter presents a novel hardware technique at the granularity of micro-instructions that does not require compiler support, and has reduced overheads. We have incorporated this technique into a commercial design flow to produce two differing processors with four security monitors, which detect well known security threats. Experiments show that our technique incurs an additional area overhead of just 9.21% and 15.4% for the two architectures. The clock period increase for both the architectures are 0.67% and 1.92%, and leakage power consumption overheads are 7.63% and 16.7%.

5.1   About this Chapter

5.1.1   Objectives

This chapter has the following primary objectives.

1. Showing that, most of the common security threats could be handled by hardware based inline monitoring.

2. Showing that the inline monitoring overheads are comparably low.

3. Presenting a security monitoring technique at a level lower than machine instructions and still at the instruction level, with better performance penalties.

5.1.2 Outline

This project presents a novel hardware integrated framework to detect security threats by performing inline* monitoring. We use a technique called the Hardware Assisted MONitoring for inline securitY checking (HARMONY) to deal with security and reliability, and show that by handling this problem at the granularity of micro-instructions we are be able to reduce the overheads to a considerable minimum.

HARMONY uses micro-instruction routines to perform runtime hardware monitoring for security. Micro-instructions are instructions which control data flow and instruction-execution sequencing in a processor at a more fundamental level than at the level of machine instructions. HARMONY does not require software instrumentation which is necessary for most of the other hardware assisted techniques. Therefore, our framework could be used without recompilation for legacy code and could be used with new applications.

In this chapter, we address the security issues of a computing system by focusing on detecting security attacks defined by the security policy of a system. We explain the checking architecture that could be deployed in any embedded or general purpose processor to ensure code integrity of an application.

We have evaluated area and delay overheads associated with the proposed architecture, for security policies to detect well known security threats. We used applications from an embedded systems benchmark suite called MiBench [115] to verify the correctness of the monitors. Hardware synthesis and simulation were performed by commercial design tools and the results demonstrate our proposed solution has minimal overheads.

Our work is inspired by the work of reliability and security engine (RSE) [192] in providing a wide range of security detection mechanisms with different monitoring techniques within the same framework. We alter the micro-instructions of critical machine instructions to perform security monitoring. In comparison to [192], our method neither needs compiler support nor an external interface to the core processor to communicate with the monitoring hardware.

5.1.3 Contributions

Contributions of this project are:

1. we show a methodology for embedding security monitoring within the micro-instructions of machine instructions;

---

*Security monitoring is performed in parallel to the regular processing. This is achieved by implementing the monitoring routines at the level of micro-instructions, which are used to form the machine instructions based on the instruction set definition

2. a hardware assisted technique that does not require compiler support and therefore could be used with legacy code as well as newly written code; and

3. an approach that does not require external interfaces between the core processor and the monitoring hardware, and therefore eliminates the need for an interface, that will depend on the type of monitoring performed.

5.1.4   Rest of this chapter

The remainder of this chapter is organised as follows. Section 5.2 presents the architecture of the inline security monitoring framework and security policies. Section 5.3 describes a systematic design methodology of the proposed solution for a given application. Results are presented in Section 5.4, future work in Section 5.5 and a summary in Section 5.6.

5.2   Monitoring Framework

In this section we give an overview of the proposed monitoring framework. We explain the architecture enrichment that the framework demands and the security policies to incorporate within this framework.

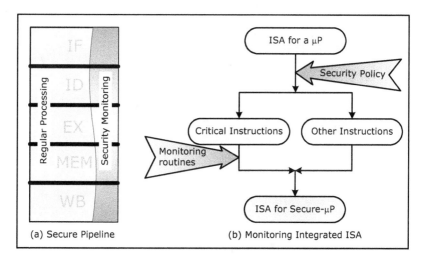

Figure 5.1: *An Overview of the Monitoring Framework*

5.2.1   Overview

As in Figure 5.1(a), security monitoring in the HARMONY framework is integrated into
the same pipeline where general purpose processing is performed. Security monitoring is
accomplished in parallel to the regular processing at the granularity of micro-instructions.
Figure 5.1(b) shows an abstract model on how the processor hardware is utilized for security monitoring. As depicted in Figure 5.1(b) the instruction set architecture (ISA) of
a particular type of micro processor is processed by considering the security policies defined by the authority. Critical instructions in the ISA, as defined by the security policies
are separated and their micro-instructions are altered by adding security monitoring routines in micro-instructions. Now the critical instructions belonging to the ISA are merged
with the other instructions to form a secure ISA. The final ISA which ensures security is
used to design the secure micro processor.

5.2.2   Embedded Micro Monitoring

As mentioned previously, monitoring routines to check insecure operations are embedded into machine instructions by the addition of extra micro-instructions, forming self-
monitoring machine instructions. A secure micro coding is more reliable, as it is performed at the architecture level and the user cannot overwrite them. They consume as little overhead as possible, since the monitoring routines are formed with micro-instruction.
When a security threat is detected one of a number of possible options can be taken. These
are:

1. terminate the current process and inform the operating system about the fault by
   returning a trap signal;

2. continue execution of the program in a safe way; or,

3. kill the insecure process.

However, analysing recovery mechanisms are beyond the scope of this project and are not
covered here.

5.2.3   Software Interface

A major advantage of our framework is that it does not require an explicit software interface. Therefore, we do not have to modify compilers and we could use this framework
with legacy as well as new applications without any changes to the existing software infrastructure. Since the architectural changes are made inline and the hardware monitoring
is performed by the same critical instructions, an explicit software interface is not needed.

5.2.4 Security Policies and Architectural Changes

Security monitoring for different threats are defined by distinctive security policies. Each of the policies may involve one or more machine instructions. The machine instructions involved in the security policies are defined as critical in our framework.

Policy I : Monitoring Return Address Overwriting

Even though there are countermeasures in software for buffer overflows, it continues to be one of the main threats in recent years [294]. Majority of buffer overflow attacks are related to overwriting the return address in the buffer to an address of a malicious function pointer by which, the control flow is transferred to the attacker. When a *call* instruction is executed the processor transfers control to the target function and upon completing the procedure, control is returned back to the instruction that follows the call instruction. In a procedure call, the last state of the processor before the procedure call is retained by storing the values in a LIFO (last in first out) buffer (as in stack). An attacker exploits the weakness of a called procedure by overwriting the return address in the stack with a value of the attacker's choice, to obtain control of the processor.

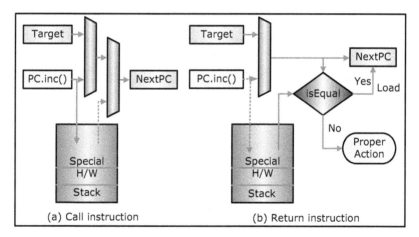

Figure 5.2: *Return Address Monitoring*

As shown in Figure 5.2, return address monitoring is used to verify the return address of a *return* instruction that is associated with the last call instruction. The solid lines in Figure 5.2 are the active lines for a particular operation. During the execution of each linked branch (*call*) instruction, a copy of the return address (next to the current PC; indicated by PC.inc()) is stored in a special hardware stack as shown in Figure 5.2(a). Then the PC is set to the target address (indicated as *NextPC* in Figure 5.2(a)). When a return address (indicated as *Target* in Figure 5.2(b)) is popped from the memory stack, it is compared against the value stored in the hardware stack using a comparator as shown in

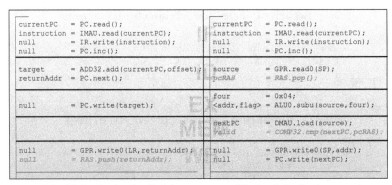

(a) call instruction (ex: b <offset>)   (b) return instruction (ex: pop pc)

Figure 5.3: *Micro Instructions for Return Address Monitoring*

Figure 5.2(b). If there is a match between the two, then the target address is set to the PC. When there is a mismatch, a signal could be sent to indicate a possible buffer overflow attack.

Figure 5.3 shows the micro-instructions of a *call* and a *return* instructions. The processor pipeline is assumed to be divided into five stages: Instruction Fetch (IF); Instruction Decode (ID); Execution (EX); Memory Read/Write (MEM); and Register Write Back (WB). Micro-instructions in ***bold-italics*** are instructions which allow embedded monitoring. These are inserted into the *call* and *return* instructions as illustrated in Figure 5.3(a) and (b).

The first four lines of micro-instructions belong to the IF stages in both *call* and *return* instructions (these four lines are identical in *call* and *return*). Among these four: the first reads current Program Counter (PC) value; the second reads the current instruction via Instruction Memory Access Unit (IMAU); the third writes the current instruction into Instruction Register (IR); and the last one increments the PC to the next instruction memory address. These four micro-instructions perform instruction fetching at the level of micro-instructions. First micro-instruction in the ID stage of the *call* instruction calculates the target branching address by adding current PC value to the offset and second micro-instruction stores the return address of this function call. EXE stage of the call instruction writes the target branching address to the PC and the first micro-instruction in WB stage writes the return address to the link register (LR). The last micro-instruction in the *call* instruction is pushing the return address into a special hardware stack called *RAS*.

The first micro-instruction in the ID stage of *return* instruction reads the stack pointer (SP) value and the second micro-instruction (a monitoring instruction) pops the return address from *RAS*. In EXE stage two micro-instructions are used to calculate the next SP (*addr*). First micro-instruction in MEM stage loaded the return address (*nextPC*) via Data Memory Access Unit (DMAU) and the second (a monitoring micro-instruction) one compares it with the one popped from *RAS*. The WB stage of the *return* instruction writes the new SP value and loads the PC with the return address. The flag from the comparison (*valid*) will be used to send a signal when a return address overwriting has been detected.

Policy II : Monitoring Unauthorized Instruction Memory Accesses

Software integrity makes sure that the program in the system is not altered or deleted by someone. By monitoring unauthorized instruction memory accesses we could be able to detect some of the unintended control flows which leads to run malicious code injected by an attacker. Memory access faults and faulty jumps are detected by storing the memory boundary values to special registers at load time. When a jump instruction is encountered, the target address is compared with the boundary values to verify that they jump to a memory location within the boundary values.

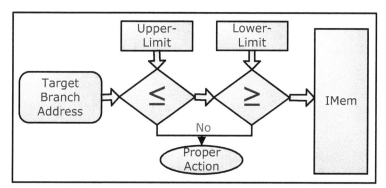

Figure 5.4: *Instruction Memory Access Monitoring*

Figure 5.4 shows how a jump/branch instruction is monitored. Such an instruction transfers control to another instruction in the instruction memory location which is within the boundary of the current program. When the target address of a jump/branch instruction is computed, it is compared with the upper and lower boundary addresses. If the target address is out of the boundary values then an unauthorized instruction memory access signal will be raised. Since this monitoring is related to security, sufficient and proper condition is covered. That is, only illegal jumps out of the boundaries are detected and illegal jumps within the boundaries are not detected.

Figure 5.5 shows the micro-instructions for a conditional *branch* instruction (*beq <const>*; i.e. branch equal). Instructions in ***bold-italics*** in Figure 5.5 belong to the monitoring routine. The micro-instructions span the first three stages of the pipeline: IF, ID and EX. Instructions belonging to the IF stage perform instruction fetching identical to the one described in Figure 5.3. In ID stage, the first two micro-instructions read the upper and lower boundaries of the instruction memory. The next three micro-instructions in the ID stage read the *FLAG* register and decide whether the condition for branching is either true or false. The last micro-instruction in ID stage sign extends the constant from the branch instruction (*beq <const>*) and stores it in a variable called *offset*.

```
currentPC    = PC.read();
instruction  = IMAU.read(currentPC);
null         = IR.write(instruction);
null         = PC.inc();
```
IF

```
upperLimit   = UPPER_IM.read();
lowerLimit   = LOWER_IM.read();
flagArray    = FLAG.read();
flagCond     = flagArray[2];
cond         = flagCond == '1';
offset       = EXT8.sign(const);
```
ID

```
target       = ADD0.adc(currentPC,offset);
valid0       = COMP32LE.cmp(upperLimit,target);
valid1       = COMP32GE.cmp(lowerLimit,target);
valid        = valid0 & valid1;
null         = [cond] PC.write(target);
```
EX

Figure 5.5: *Micro-instructions for Instruction Memory Access Monitoring*

The first micro-instruction in the EX stage in Figure 5.5 adds *offset* from ID stage to *currentPC* from IF stage to calculate the target address for branching and stores it in *target*. The second and third lines in EX stage compare the *target* against the upper and and lower boundaries read in ID stage. Fourth line combines the outcomes of lines two and three and stores the result in *valid*, which will be used to send a signal when there is a possible unauthorized instruction memory access. The last micro-instruction in EX stage writes the *target* to the PC if the value in *cond* is true, so that a conditional branch will be performed. Identical micro-instruction monitoring routines will be embedded into the other branch instructions in the ISA.

Policy III: Monitoring Unauthorized Data Memory Accesses

Data integrity ensures that data in an embedded system has not been altered or deleted by someone. Some of these attacks could be detected by monitoring unwanted data accesses. Data memory access faults are detected by storing the addresses of data memory boundaries to special registers at load time. When a *load/store* instruction is encountered, the target address is compared with the boundary values to verify that they access a memory location within the boundary values.

Figure 5.6 shows how a load/store instruction is monitored. Such an instruction accesses data memory to read/write a data value. When the target address of a load/store instruction is computed, it is compared with an upper and a lower boundary addresses. If the target

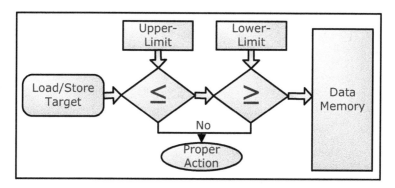

Figure 5.6: *Data Memory Access Monitoring*

address is out of the boundary values then an unauthorized data memory access signal
will be raised.

```
currentPC     = PC.read();
instruction   = IMAU.read(currentPC);                        IF
null          = IR.write(instruction);
null          = PC.inc();

source0       = currentPC;
source1       = EXT8.zero(const);                            ID
upperLimit    = UPPER_DM.read();
lowerLimit    = LOWER_DM.read();

<addr,flag>   = ALU0.add(source0, source1);
valid0        = COMP32LE.cmp(upperLimit,addr);               EX
valid1        = COMP32GE.cmp(lowerLimit,addr);
valid         = valid0 & valid1;

result        = DMAU.load(addr);                             MEM

null          = GPR.write0(rd,result);                       WB
```

Figure 5.7: *Micro-instructions for Data Memory Access Monitoring*

Figure 5.7 shows the micro-instructions for a PC relative *load* instruction (*ldr <rd>*
<const>). Micro-instructions in ***bold-italics*** in Figure 5.7 belong to the monitoring rou-

tine. Micro-instructions for this *load* instruction span to all five stages of the pipeline: IF, ID, EX, MEM and WB. Micro-instructions belonging to the IF stage perform instruction fetching in similar manner to the one described in Figure 5.3. In ID stage, the first micro-instruction copies current PC value to *source0*. The second micro-instruction extends the constant *const* and stores the outcome in *source1*. The last two micro-instructions in ID stage read the upper and lower boundaries of the data memory.

The first micro-instruction in the EX stage in Figure 5.5 adds *source0* and *source1* from ID stage to calculate the target address for loading the data and stores this address in *addr*. The second and third lines in EX stage compare this *addr* against the upper and lower boundaries read in ID stage. Fourth line combines the outcomes of lines two and three and stores the result in *valid*, which will be used to send a signal when there is a possible unauthorized data memory access.

The micro-instruction in MEM stage loads the data value from data memory to *result* via DMAU, which is written back to the register denoted by *rd* at the WB stage.

Policy IV : Monitoring Fault Injection into the Instruction Memory

Changes in environmental conditions and external parameters such as temperature, humidity, radiation, and supply voltage are changed to induce faults in an embedded system component, particularly the memory. Eventually this type of attack will make the system non-reliable and therefore the attacker changes a security threat into a reliability problem. Historically these are called side channel attacks, and are performed on very low-end embedded systems such as smart cards [119, 141, 144, 220]. But recently these attacks are becoming more and more robust such that they could be a security threat to any embedded processors. The following are the main methods in which fault injection attacks compromise the security of an embedded system [227]:

1. **availability:** disable the availability of a device by occupying it with unwanted data or by damaging it;

2. **integrity:** damaging the data and application stored in storage devices; and

3. **privacy:** gaining access to secure data by breaking various cryptographic schemes by taking advantage of random faults as described in [29].

Fault injection into the instruction memory is handled by the fault injection monitor as depicted in Figure 5.8. A duplicate copy of the binary instructions is stored in a different location of the memory. This duplication could be done on all the binary instructions or part of the binary instructions according to the level of availability requirements. When an instruction is fetched by the ordinary IMAU, a duplicate copy of the instruction is fetched by a special IMAU and before the instruction is written into IR, the original is compared against the duplicate copy. A mismatch in the comparison may indicate a fault injection attack. Note that this same monitor could be enhanced to monitor code injection attacks. Since the injected code is going to overwrite the original binary, if the duplicate copy of

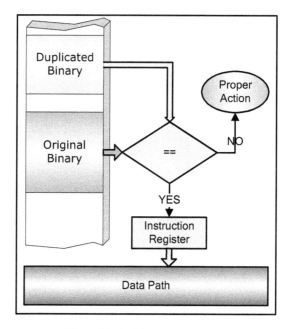

Figure 5.8: *Fault Injection Monitoring*

the binary is stored in a secure memory segment, the comparison at the instruction fetch stage will identify the mismatch caused by the code injection attack.

```
currentPC    = PC.read();
instruction  = IMAU.read(currentPC);
spInst       = spIMAU.read(currentPC + offset);
valid        = COMP16.cmp(instruction,spInst);
null         = IR.write(inst);
null         = PC.inc();
```

Figure 5.9: *Micro-instructions for Fault Injection Monitoring*

Figure 5.9 shows micro-instructions (IF stage) of all the critical instructions, defined by fault injection monitoring security policy. The micro-instructions in Figure 5.9 assumes the whole binary is duplicated. Micro-instructions in ***bold-italics*** are belong to the monitoring routine and the others do instruction fetching identical to the one described in Figure 5.3. The first micro-instruction in ***bold-italics*** loads the duplicate copy of the instruction from an offset memory location to *spInst* and the second micro-instruction

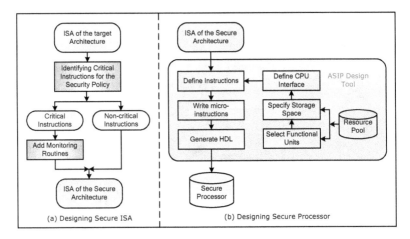

Figure 5.10: *Design Methodology*

compares it with the one loaded in the ordinary fetching process. A mismatch in the comparison will be used to signal a possible fault injection into the instruction memory.

5.3 Architectural Design Flow

In this section, an overview of the proposed design flow for the monitoring architecture is provided. Without losing generality of our technique, we use an automatic processor design tool to implement our design in hardware. This automatic design tool is used to design Application Specific Instruction-set Processors (ASIPs) [91, 92], custom designed for applications or application domains.

Figure 5.10(a) describes the design of a secure ISA from a given ISA for a target Architecture. Critical instructions, for which the micro-instructions have to be altered are identified by using the defined security policies. These critical instructions are altered by adding monitoring routines as described by the security policies. Critical instructions with the amended micro-instructions are merged with non-critical instructions to form secure ISA for the target processor.

Figure 5.10(b) depicts the hardware design flow that starts from a secure ISA (as described by Figure 5.10(a)) and produces a Hardware Description Language (HDL) implementation of the secure processor. First, from a pool of resources functional units and storage spaces required for a processor are selected. Then the interfaces of the CPU to the memory and other devices are defined. Machine instructions for the target ISA are defined at this stage. Security monitoring routines are inserted into the logic, while writing micro-instructions for each machine instruction. Micro-instructions are written with due consideration to the pipeline stages of the target processor architecture. Finally, the secure processor implementation in an HDL is generated for hardware synthesis and simulation.

5.4 Experimental Results

In this section, we present the hardware area overheads incurred by the proposed architecture, as well as the impact of the technique on performance. For the purpose of experiments, we have used two different instruction sets to implement our secure processors: (1)
ARM Thumb™ [287, 288] and (2) PISA [47] (as implemented in SimpleScalar® tool-set
[37]). Applications from MiBench benchmark suite are used in the experiment.

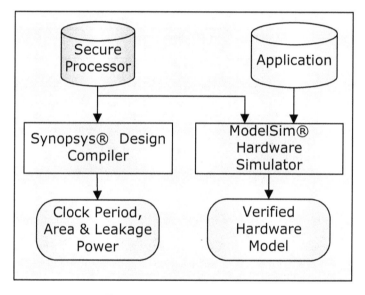

Figure 5.11: *Evaluation Methodology*

Evaluating the Overheads

To evaluate the methodology, applications from MiBench benchmark suite are taken and
compiled with the GNU/GCC® [289] cross compilers for ARM and PISA instruction sets.
An automatic ASIP design tool, ASIP Meister [293] is used to generate the VHDL description of the secure processors as described in Section 5.3. Two different sets of
ISAs are used to generate secure processors, one for ARM Thumb™ and the other for
PISA instruction sets. The outputs of ASIP Meister are the VHDL models for simulation
and synthesis of the secure processor. Synthesis models are used with Synopsys Design
Compiler® [295] (as shown in Figure 5.11) to obtain the area, clock period and leakage
power overheads. Taiwan Semiconductor Manufacturing Company's (TSMC) 90nm core
library is used for synthesis with typical conditions enabled.

ModelSim [290] hardware simulator is used with application binaries to verify the correctness of our monitoring hardware (as shown in Figure 5.11). Different applications are

run with known inputs and outputs to verify the correctness of the hardware. Further, operation of each instructions of the ISA are tracked on the wave window of the simulator to verify their operations. For example, when a particular instruction is fetched, how the data flow through different components of the datapath is observed for its correctness. Further, to test each monitor, the particular scenario violating the security policy is generated and the reaction of the hardware for this scenario is observed for its correctness.

Monitor	ISA	Clock Period (ns)	Area (1000 gates)	Leakage Power (μW)
MonNo	ARM Thumb [TM]	4.68	104	186
	PISA	16.94	228	485
MonFIIM	ARM Thumb [TM]	4.77	104	187
	PISA	17.05	230	489
MonRAO	ARM Thumb [TM]	4.70	116	206
	PISA	16.98	243	511
MonUIMA	ARM Thumb [TM]	4.68	107	193
	PISA	16.94	231	493
MonUDMA	ARM Thumb [TM]	4.76	106	190
	PISA	17.03	232	493
MonAll	ARM Thumb [TM]	4.77	120	217
	PISA	17.05	249	522

Table 5.1: *Clock Period, Area and Leakage Power Overheads of HARMONY*

Table 5.1 shows the clock period, area and leakage power comparisons of our processors due to extra monitoring logic implementation. Different monitors as described by the security policies in Section 5.2.4, are implemented separately and collectively and the results are tabulated. The rows indicated by *MonNO* shows design parameters of the original processor without any monitoring enabled. Four pair of processors are synthesised to include the four different security policies described in Section 5.2.4 and they are:

1. *MonFIIM*: Monitoring Fault Injection into the Instruction Memory (as in Policy IV, Section 5.2.4);

2. *MonRAO* : Monitoring Return Address Overwriting (as in Policy I, Section 5.2.4);

3. *MonUIMA*: Monitoring Unauthorized Instruction Memory Accesses (as in Policy II, Section 5.2.4); and

4. *MonUDMA*: Monitoring Unauthorized Data Memory Accesses (as in Policy III, Section 5.2.4).

The rows indicated by *MonAll* represents processors designed by including all the four security policies as described above.

The clock period of a processor indicates the maximum time needed for the data to pass the critical path of the pipeline stages in a pipelined processor. In Table 5.1, clock periods

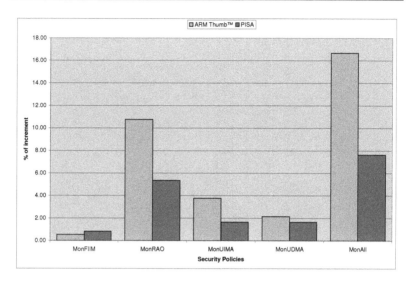

Figure 5.12: *Percentage of Clock Period Overhead*

for processors with no monitoring enabled are 4.67ns for ARM Thumb™ and 16.94s for PISA processors. When all four monitors are enabled (*MonAll*) the clock periods are 4.77ns (ARM Thumb™) and 17.05ns (PISA). Four pairs of processors where one security monitoring is enabled on each have clock periods between the minimum (*MonNo*) and the maximum (*MonAll* values, except *MonFIIM* which also has the maximum clock period.

It is obvious that an increase in logic will cost more area. As shown in Table 5.1, while the processor pair without monitoring require 104 (*Arm Thumb™*) and 228 (*PISA*) thousand gates, the pair where all four monitors enabled require 120 (*Arm Thumb™*) and 249 (*PISA*) thousand gatess respectively. The leakage power consumed by a processor will increase with the number of cells used in the processor hardware. Table 5.1 expresses this fact, where the leakage power consumptions for the processor pair without monitors are 186μW and 485μW, the leakage power figures with all four monitors enabled are 217μW and 522μW.

$$MonAll_{CP} = MAX(MonFIIM_{CP}, MonRAO_{CP}, MonUIMA_{CP}, MonUDMA_{CP}) \quad (5.1)$$

Figure 5.12 depicts percentage increments in clock periods with respect to the processor pair (ARM Thumb™ and PISA) with no monitoring enabled. The percentage of increase ranges from no changes in *MonUIMA* pair to 1.92% and 0.65% in *MonAll* pair. From the chart it is evident that the clock period increases when all the monitors are enabled is equal to the maximum clock period increase among the monitors, *MonFIMM* in the chart.

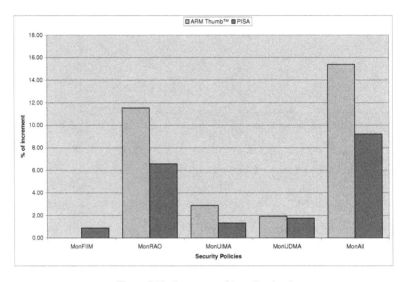

Figure 5.13: *Percentage of Area Overhead*

Equation (5.1) represents this fact and the notation X_{CP} in the same equation represents the clock period increase of monitoring policy X.

$$MonAll_{Ar} = \sum(MonFIIM_{Ar}, MonRAO_{Ar}, MonUIMA_{Ar}, MonUDMA_{Ar}) - SC_{Ar} \quad (5.2)$$

Figure 5.13 depicts the percentage of area overheads with respect to the processor pair with no monitoring enabled. The total area overhead when all the monitors are enabled should be the accumulated sum of individual monitors. But, due to the sharing among the monitoring hardware the total overhead will be given by Equation(5.2), where the notation X_{Ar} represents the percentage of area overhead of X with respect to the hardware without monitoring. SC_{Ar} in the equation stands for the area of the shared components.

$$MonAll_{LP} = \sum(MonFIIM_{LP}, MonRAO_{LP}, MonUIMA_{LP}, MonUDMA_{LP}) - SC_{LP} \quad (5.3)$$

Figure 5.14 depicts the percentage of leakage power overheads with respect to the processor pair with no monitoring enabled. The total leakage power consumption overhead when all the monitors are enabled will be the accumulation of leakage power consumption of each monitors implemented separately. But, as discussed in the section dealing with area overheads, component sharing will favor a little deduction in the total leakage power consumption. This is expressed in Equation (5.3), where the notion X_{LP} represents the leakage power overhead of the processor X with respect to the leakage power consump-

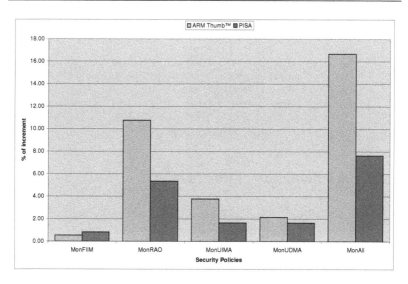

Figure 5.14: *Percentage of Leakage Power Overhead*

tion of the processor without monitoring. SC_{LP} in the equation stands for the leakage power of the shared components.

In [300], Vetteth has implemented the RSE infrastructure (the external interface between the processor core and the monitoring hardware) and two of the RSE modules using a hardware description language. He has also evaluated the performance and area overheads by synthesizing the hardware in Xilinx® FPGA, VirtexE1000-6. For a double issue DLX processor [85], an exhaustive RSE infrastructure (only the interface) itself occupies 11% extra area, and both the RSE modules occupy 4% extra area which adds up to 15% area overhead. A customized version of the RSE infrastructure with the two modules costs 6% extra area. Our framework with four monitors occupy an area overhead of 9% where *MonRAO* (which was not one of the two modules in Vetteth's implementation) incurs almost 7% of the area overhead due to an additional hardware stack implemented in the monitor.

The clock period overhead for Vetteth's implementation of two RSE modules was 5.5%; however, the overhead for four modules in our framework (PISA instruction set) was a negligible 0.65%. Vetteth has not evaluated the clock cycle penalty due to the software instrumentation. Our framework does not incur clock cycle overhead as we do not add extra instructions into the instruction stream. A significant improvement to note is that if our monitors were to be implemented in RSE, then compiler support would have been essential. As it stands now, we do not require compiler support and therefore we support security monitoring of legacy code as well as new applications.

5.5   Future Work

Even though the security policies in our framework are defined systematically, it is possible to use formal methods to define the policies and automatically generate microinstruction routines for monitoring based on the defined policies. We identify this as a future work. Futher, this project has not considered and implemented the possible recovery mechanisms when a security threat is detected. Recovery integration is another future work, which is considered worth analysing.

5.6   Summary of this chapter

In this chapter, we have presented a simple hardware-assisted framework for runtime inline security monitoring at the granularity of micro-instructions. We have defined a number of security policies and shown how they could be implemented in our framework with negligible performance and very little area overhead. We have elaborated an automatic technique to design our framework using an automatic processor design tool. Hardware overheads, clock period changes and leakage power consumption changes are evaluated and reported. The results show that these overheads are small and reasonable. Our study reveals that the proposed HARMONY framework is capable of handling any inline security monitoring with minimal overheads. We conclude this chapter by saying that our framework is useful in answering the increasing security and reliability concerns in computer systems.

Chapter 6

Encrypted Basic Block Check-summing for Reliable and Secure Embedded Processors

... 'Speak English!' said the Eaglet. 'I don't know the meaning of half those long words, and, what's more, I don't believe you do either!'

— Lewis Carroll, *Alice's Adventures in Wonderland*

Security and reliability in processor based systems are concerns requiring adroit solutions. Security is often compromised by code injection attacks, jeopardizing even 'trusted software'. Reliability is of concern, where unintended code is executed in modern processors with ever smaller feature sizes and low voltage swings causing bit flips. Countermeasures by software-only approaches increase code size and therefore reduce performance. Hardware assisted approaches use additional hardware monitors and therefore not scalable. In this chapter we propose a novel scalable hardware/software technique to reduce overheads. Experiments show that our technique incurs an additional hardware overhead of 5.03% and clock period increase of 0.06%. Average clock cycle and code size overheads are 3.66% and 10.6% for five industry standard application benchmarks. Fault injection study reveals full coverage of bit burst detection.

6.1 About this Chapter

6.1.1 Objectives

This chapter has the following objectives:

1. providing a software/hardware technique to detect basic block code integrity violations and therefore code injection attacks;

2. showing that the same technique will also capture instruction memory bitflips; and

3. evaluating the related overheads, fault coverage, and fault detection latency of the proposed technique.

6.1.2 Outline

This chapter presents a novel hardware integrated technique to deal with code injection attacks and bit flips caused by transient faults. That is, we propose to add security detection mechanism during the design phase of a processor, rather than adding it as a feature later in the process. For the first time, we use a technique called the Integrated Monitoring for Processor REliability and Security (IMPRES) to deal with this and show that by handling this problem at the granularity of micro-instructions (MIs) and making security a design parameter, we will be able to reduce the overheads. Additionally, we use fault injection experiments to show that IMPRES has full fault coverage for bit flips in instruction memory.

IMPRES uses micro-instruction routines and software instrumentation to perform run-time hardware monitoring for reliability and security. Micro-instructions are instructions which control data flow and instruction-execution sequencing in a processor at a more fundamental level than at the level of machine instructions. The software instrumentation performed in IMPRES is minimal compared to software only approaches, because it is used only as an interface between check points and micro-instruction routines.

In this chapter, we address security and reliability of a computing system by focusing on code injection attacks and transient bit flips. micro-instructions for monitoring are embedded into machine instructions and therefore the extra burden of performing security checks are evenly distributed. We explain the IMPRES framework and design methodologies that could be deployed in any embedded processor to ensure code integrity of an application.

We evaluate the area and delay overheads associated with the proposed architecture, and performance impact using applications from an embedded systems benchmark suite called MiBench [115]. Hardware synthesis and simulation are performed by commercial design tools and the results demonstrate our proposed solution has minimal overheads.

6.1.3 Contributions

The ultimate goal of security attacks is to gain control of the system and destroy system integrity by altering information which is in the form of software and data. Embedded Micro Monitoring (EMM) [222] is an architectural framework that uses micro-instruction routines to perform in-line security monitoring. EMM performs the checking without modifying the application program, but by only changing the micro-instructions for selected machine instructions. EMM provides support for reliability at a cost of doubling the memory [221], and partial support for software integrity attacks [222].

Described for the first time here by IMPRES is a generic, scalable, low overhead method to both improve reliability and protect against code injection attacks. Software-only methods only detect bit flips and are susceptible to code injection attacks [319]. Hardware methods use tables [11] or watchdog processor's memory [155, 162] which are neither scalable nor relocatable. Ours overcomes these shortcomings. Additionally, by the use of encryption, we protect against an attacker changing a whole basic block with the checksum, to gain access to the executable.

Thus our **contributions** are:

1. we have demonstrated a *simple* hardware/software method for monitoring code injection attacks which requires only a rudimentary software analysis;

2. this monitor is also capable of picking up transient faults such as flips. We have used a fault injection engine to show the fault coverage of this technique; and

3. produces code which is relocatable allowing the use of this technique with an operating system.

We believe that the above contributions will make practical the deployment of software integrity checkers for real applications.

The IMPRES framework detects code injection attacks and bit flips in instruction memory. IMPRES will neither detect other security threats, such as second order code injection attacks (malicious applications loaded and executed by unknowledgeable users) nor reliability problems, such as bit flips in data memory. A solution for second order code injection attack is described in [114] and bit flips in data memory could be protected by error-correcting-code. Furthermore, a full control over the processor design process is assumed. Our decision to dictate IMPRES as a design time solution supports this assumption.

6.1.4   Rest of this chapter

The remainder of this chapter is organized as follows. Section 6.2 presents the proposed IMPRES framework that includes an overview of the basic block code integrity and the basic block integrity violation model. Section 6.3 describes systematic software and hardware methodologies to design the proposed solution for a given application. Experimental setup and evaluation are in Section 6.4 and results in Section 6.5. Section 6.6 details fault injection analysis and Section 6.7 proposes some future directions for this project. Finally, Section 6.8 summarizes IMPRES project.

6.2   IMPRES Framework

In this section, we give an overview of our monitoring framework, IMPRES and discuss the basic block code integrity violation model for IMPRES. We argue that the runtime code integrity could be fully preserved if all the basic blocks of an application program are ensured to be intact. Thus only performing basic block integrity checking is sufficient to ensure software integrity.

6.2.1   Basic Block Code Integrity

The proposed basic block integrity checker incorporates the following three tasks:

1. identifying basic blocks and calculating and assigning checksums for each basic block statically;

2. encrypting the checksums with a secret hardware key at load time, generating encrypted checksums; and

3. re-calculating the encrypted checksums at runtime and comparing the encrypted checksums with loaded values using hardware enhancements

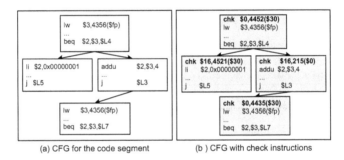

(a) CFG for the code segment   (b) CFG with check instructions

Figure 6.1: *Statically Performed Software Instrumentation*

Figure 6.1 illustrates the software instrumentation for a given code segment. A code segment grouped into basic blocks, as in Figure 6.1(a) based on the control flow of the code segment are processed separately to calculate checksums based on the instructions of each block. The calculated checksum is then inserted at the beginning of each basic block using a special instruction (*chk* instruction as in Figure 6.1(c)). Details about the software instrumentation process is discussed in Section 6.3.1. Even though, the diagrams in Figure 6.1 depict control flow graphs (CFGs) of a code segment, it is sufficient to identify control flow instructions (CFIs) and the addresses (labels) to which the CFIs will potentially branch/jump. Therefore, it is unnecessary to formally generate CFGs to perform basic block check-summing.

Figure 6.2 depicts how a basic block with a checksum is securely loaded into memory, and how a hardware integrated monitor is used to detect code integrity violations at runtime. A basic block with calculated checksum is loaded using a secure loader. While loading instructions, the loader will use a hardware key to encrypt the calculated checksum into an encrypted checksum, using an encryption algorithm. Data Encryption Standard with a 56-bit key, known as DES56 is used in IMPRES. However, it is possible to incorporate

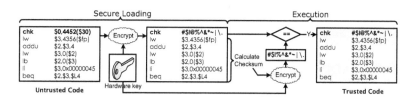

Figure 6.2: *Secure Loading and Execution*

any other data encryption algorithms such as 3DES (tripple DES) or AES (Advanced Encryption Standard) into IMPRES when a need arises. Any encryption standard could be used as long as you are able to implement a hardware core of that encryption standard.

Right half of Figure 6.2 depicts how a code integrity violation is captured at runtime. The first instruction of a loaded basic block is a *chk* instruction that carries the encrypted checksum for the corresponding basic block. When an instruction of this kind is fetched, the encrypted checksum is loaded into a special register (*eChkSum*). Checksum for each basic block is incrementally re-calculated at runtime, while instructions belonging to the basic block are executed and is stored in another special register (*iChkSum*). The incremental re-calculation is achieved using micro-instructions integrated into each machine instruction. Last instruction of each basic block is a CFI and if not, one is inserted (if it is not present at the end of a basic block) micro-instructions for CFIs are altered such that they will (a) encrypt the incrementally stored checksum with the same hardware key used during loading and (b) compare the result against the one loaded from *chk* instruction. A mismatch in the comparison will indicate a code integrity violation and generate a *SIGCKSM* signal.

Apart from encrypted checksums, a special single bit flag (*fBB*) is used to capture code integrity violations those escape the encrypted checksum technique. When a program is loaded and a CFI is executed, *fBB* is set. When a non CFI is executed *fBB* is cleared. When a *chk* instruction is executed and if *fBB* is not set (this occurs when the CFI of the last basic block is not executed) then a 'no control flow instruction' error (*SIGNCFI*) is signaled.

Our technique hugely differs from other checksum- or hash-based software integrity checking techniques, where hashing is performed at the beginning or end of basic blocks and therefore accumulates the workload to particular points in the program flow. IMPRES distributes the overhead to all the instructions, thus the total hardware related overhead is reduced to an amount that is negligible. Encryption algorithms are complex and incur higher overheads [34], while calculating checksums are less complex. Therefore we have used check-summing together with encryption to perform just a few encryptions while ensuring that the technique is still secure.

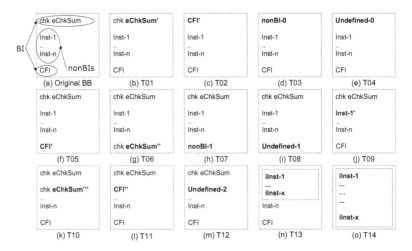

Figure 6.3: *Basic Block Code Integrity Violations*

6.2.2   Basic Block Code Integrity Violation Model

A number of determinants could be considered when choosing program properties to be monitored. Importantly the properties chosen should clearly indicate when a violation of behavior occurs. As we have already discussed, we will only be ensuring integrity of basic blocks (BB).

After software instrumentation, each BB will start with a *chk* instruction and end with a CFI. We will call the *chk* and the CFI, the boundary instructions (BI) and all others in each BB non boundary instructions (nonBI). If we can assure that each and every BB in an application is intact, then we can safely assume that the code integrity is enforced for the whole application. We classify the possible code integrity violations due to code corruption or code injection attack under different categories as detailed in Figure 6.3 and summarized in Table 6.1.

In Figure 6.3, block (a) depicts an original, unaltered BB and blocks (b)-(o) depict all the possible code integrity violations, which could take place in a BB. Blocks (b)-(e) show the cases when the *chk* instruction in the original BB is changed into a *chk* with a different encrypted checksum (b); a CFI (c); a nonBI (d); or a binary that is non decodable (e) - no opcode exist, indicated by *Undefined*. Blocks (f)-(i) depict the cases, when the CFI in the original BB is turned into another CFI (f); a *chk* (g); a nonBI (h); or an undefined instruction (i). Blocks (j)-(m) depict the cases, when any one of the nonBIs in the original is replaced by another nonBI (j); a *chk* (k); a CFI (l); or an undefined instruction (m). In Figure 6.3, (n) represents the case, when more than one instructions from the begining of BB are corrupted and (o) represents the whole BB being corrupted.

We claim that all the other combinations of basic block code integrity violations, which are not shown in Figure 6.3 are subsets of the cases presented in the same figure and

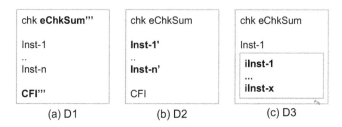

chk **eChkSum'''**	chk eChkSum	chk eChkSum
Inst-1	**Inst-1'**	Inst-1
..	..	**iInst-1**
Inst-n	**Inst-n'**	...
CFI'''	CFI	**iInst-x**
(a) D1	(b) D2	(c) D3

Figure 6.4: *Basic Block Code Integrity Violations - Duplicate Cases*

therefore it is sufficient to consider only the presented cases. For example in Figure 6.4, block (a) is a subset of Figure 6.3(b) as the *chk* instructions' checksums are changed in both the scenarios and both will result in checksum mismatches. Figure 6.4(b) is a subset of block (j) in Figure 6.3. This is becuase changing more than one nonBIs into other nonBIs will have the same effect as changing one nonBI into another nonBI, and both generate checksum mismatches at the end of the BB. The block (c) in Figure 6.4 is a subset of the block (o) in Figure 6.3, since the *chk* instruction and other nonBIs in the former block will not effect the way how the block perform checking, which is similar in the later.

Type	Original	Changed	Error Signal
T01	chk	checksum	SIGCKSM
T02	chk	CFI	SIGCKSM
T03	chk	nonBI	SIGCKSM
T04	chk	undefined	SIGSYSM
T05	CFI	another CFI	SIGCKSM
T06	CFI	chk	SIGNCFI
T07	CFI	nonBI	SIGNCFI
T08	CFI	undefined	SIGSYSM
T09	nonBI	nonBI	SIGCKSM
T10	nonBI	chk	SIGNCFI
T11	nonBI	CFI	SIGCKSM
T12	nonBI	undefined	SIGSYSM
T13	chk & nonBIs	any insts.	SIG(CKSM/NCFI)
T14	whole BB	any insts.	SIG(CKSM/NCFI)

Table 6.1: *Different Types of Code Integrity Violations*

The first column in Table 6.1 names different types of code integrity violations those matches the illustrations in Figure 6.3. Column two represents the instructions that are going to be corrupted in the original BB and column three, the injected instructions. The fourth column in Table 6.1 shows the error signal raised for each code integrity violation.

As shown in Table 6.1, types T01, T02, T03, T05, T09 and T11 violations raise *SIGCKSM* signals. When the loaded encrypted checksum of the *chk* instruction is changed, as in T01, the checksum comparison at the end of the BB will generate a mismatch and a *SIGCKSM* signal. When the *chk* instruction is converted into a CFI (as in T02), the inserted CFI itself will perform a comparison and a checksum mismatch will occur and this will result in a *SIGCKSM* signal being raised. When the *chk* instruction is changed into a nonBI (as in type T03), the CFI at the end of the BB will raise a *SIGCKSM* signal. When the *chk* instruction is changed into an undefined instruction as in type T04, a *SIGSYSM* signal (system error) will be raised as this instruction is not decodable. A violation of type T05 will result in a faulty runtime checksum as the check-summing includes the CFI. This will be detected when the CFI at the end of the BB is executed and a *SIGCKSM* signal will be raised. In T09, the runtime encrypted checksum is going to be errornous as it includes the nonBI that is changed and therefore will not match the loaded value. Therefore, for type T09, a *SIGCKSM* signal will be raised. In a violation of type T11, the checksum comparison is going to be performed by the corrupted instruction and this will result in a comparison mismatch and therefore a *SIGCKSM* signal will be raised.

Code integrity violations of type T06, T07 and T10 are going to generate *SIGNCFI* signals. In T06, the CFI is turned into a *chk* and therefore the *chk* will be executed without a CFI being executed and this will result in a *SIGNCFI* error. A violation of type T07 will result in an execution of a *chk* instruction (from the next BB) without the last CFI being executed in the current BB. Therefore, a *SIGNCFI* signal will be generated. When a nonBI is changed into a *chk* instruction (as in T10) there are going to be two consecutive *chk* instructions in the program flow and this is going to generate a *SIGNCFI* signal.

When any of the instructions in a BB is converted into an undefined instruction, the system will not be able to decode the converted instruction. This will result in system errors which are denoted by *SIGSYSM* signals. This is why BB code integrity violations T04, T08 and T12 generate *SIGSYSM* signals.

Type T13 and T14 occur, when more than one instructions including the *chk* instruction (but not the CFI) in a BB are replaced by other instructions and the whole BB is replaced by other instructions respectively. The injected instructions might contain:

SET1: one or more CFIs; or

SET2: one *chk* instruction and no CFIs; or

SET3: more than one *chk* instructions and no CFIs; or

SET4: only nonBIs.

In T13, when the injected code is represented by *SET1*, a checksum mismatch will occur, as the first CFI in *SET1* is going to perform a checksum comparison. This will result in a *SIGCKSM* signal. When the injected code is represented by *SET2* the *chk* in *SET2* will load an invalid checksum and therefore the CFI at the end of BB will generate a *SIGCKSM* signal. When the injected code is represented by *SET3*, the second *chk* instruction in *SET3* will generate a *SIGNCFI* signal as two *chk* instructions are executed without a CFI in between them. When the injected code is represented by *SET4* the CFI at the end of the BB will generate a *SIGCKSM* signal.

In type T14, the first CFI (for *SET1*) in the injected code will raise a *SIGCKSM* signal. When the injected code is represented by *SET2*, the *chk* in the following BB will raise a

SIGNCFI signal. As indicated by *SET3*, when there are more than one *chk* instructions in the injected code the second *chk* instruction will raise a *SIGNCFI* signal. When the injected code is represented by *SET4*, the *chk* instruction in the following genuine BB will generate a *SIGNCFI* signal as the *chk* in the genuine BB is not just preceded by a CFI.

A fake injected BB to act as genuine, it should contain proper BIs and in particular the *chk* instruction with a valid encrypted checksum. However, generating a valid *chk* instruction is prevented by the encryption technique described in Figure 6.2. Since it is impossible for an attacker to get hold of the randomly generated secret hardware key, the intruder will not be able to generate a valid BB with malicious code. In the worst case, if an attacker has managed to crack the hardware key for a particular processor, it is impossible to perform a mass attack as each processor will have random hardware keys, which cannot be guessed.

6.3   Design Flow

In this section, an overview of the proposed design flow for the IMPRES architecture is provided. First, the design of a software interface that allows the applications to interact with the architectural enhancement is described, and then the design of the architectural enhancement itself is discussed.

6.3.1   Software Design

Figure 6.5 describes two alternative design arrangements for the interface between an application program and monitoring hardware. It is worth noting that check instructions inserted at the beginning of BBs and micro-instructions embedded into machine instructions serve as the interface between software and hardware. The left half of Figure 6.5 depicts the compile time software instrumentation process, where the source code of an application is compiled by the front end of a compiler and the assembly code for the target Instruction Set Architecture (ISA) is produced. Then a software parser is used to instrument the assembly code. Finally, the instrumented assembly code is assembled and linked using the back-end of the same compiler to generate instrumented binary (*iBinary*) which runs on the target architecture (*IMPRES* hardware).

The right half of Figure 6.5 depicts a binary code specialization technique, called binary rewriting, where either a newly compiled binary or a legacy code is rewritten by a binary rewriter. The task of the rewriter and the compile time parser are identical, even though the former uses assembly code and the later uses binary code. When a binary rewriter is used, the output from the rewriter, which is the instrumented binary will be used to run the application on our target architecture.

Furthermore, the loader of the same compiler tool kit is modified to integrate secure loading of the instrumented binary (*iBinary*) as described in Figure 6.2.

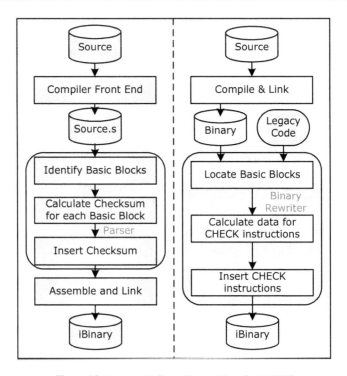

Figure 6.5: *Proposed Software Design Flows for IMPRES*

Tasks performed either by the compile time parser or the binary rewriter are:

1. identifying basic blocks;

2. calculating checksums for each basic block based on the instruction sequence of each basic block;

3. forming check instructions by inserting the checksums calculated in tasks 2 above; and

4. inserting the *chk* instructions either into the assembly or binary instruction stream based on whether the compile time parser or the binary rewriter is used.

Inserting signatures at non uniform intervals into an application binary and using these signatures at runtime to perform hardware checks has never been possible without the compiler support [183]. Since begining of each basic block (and therefore inserted check instructions) are not going to be in regular intervals in an application program, we use a compile time parser to instrument our applications as shown in the left half of Figure 6.5. However, we also propose a binary rewriting technique [284] to perform software instrumentation as depicted at the right half of Figure 6.5.

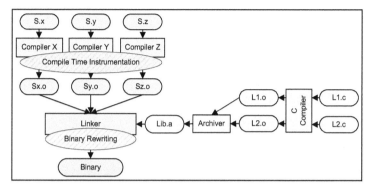

Figure 6.6: *Compile Time Instrumentation versus Binary Rewriting*

Figure 6.6 depicts the positions where compile time instrumentation and binary rewriting endure in a programming tool chain. Compile time instrumentation will only be capable of asserting checks within the user code and not functions included from library archives. But, on the other hand using binary rewriting it is possible to assert checks not only in user functions but also library functions. We are indeed assuming that all code constituting a program is available at link time, or in other words, that the programs are statically bound. This assumption is not valid for dynamic languages such as Java, where, because of reflection, the complete program may not be available. Dynamic shared libraries, of which the code is only available when a program is loaded into memory also violate this assumption. However, performing the instrumentation with binary rewriting will not only insert check instructions for library functions, but also enables instrumentation of legacy binary code. Furthermore, a binary rewriter will not require compiler modifications and therefore the same compilers could still be used without any changes.

6.3.2   Architectural Design

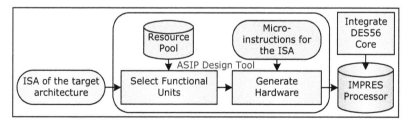

Figure 6.7: *Architecture Design Flow for IMPRES*

Without losing generality of our technique, we use an automatic processor design tool to implement IMPRES in hardware. This automatic design tool is used to design Application Specific Instruction-set Processors (ASIPs), custom designed processors for applications or application domains. Figure 6.7 describes this design process. The ASIP design tool

allows us to write micro-instructions for each machine instruction and to add new machine instructions for particular ISA. We have used this feature to add a new machine
instruction called *chk* as mentioned in Figure 6.2. We amend all CFIs (by modifying their
micro-instructions) to perform encryption and comparison as illustrated in Figure 6.2.
We also modify micro-instructions for nonBI instructions so that they will perform instruction check-summing. It should be noted that, even though we describe the hardware
improvements as micro-instruction amendments, the same could be realised as functional
additions/amendments to the hardware.

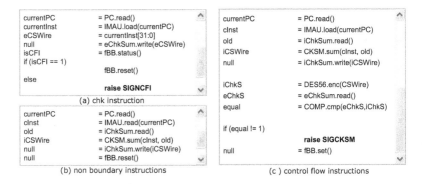

Figure 6.8: *Implanted Micro-instructions for Monitoring in IMPRES*

Figure 6.8 depicts micro-instructions inserted for monitoring in different machine instructions. Figure 6.8(a) shows micro-instructions implanted into the *chk* instruction. Encrypted checksum loaded via the *chk* instruction (*currentInst* in the figure) is stored in a
special register called *eChkSum* (lines three and four in Figure 6.8(a)). The last five lines
of Figure 6.8(a) show how flag *fBB* is used to capture the execution of consecutive *chk*
instructions. Micro-instructions added inside all non-boundary instructions to do incremental check-summing is shown in Figure 6.8(b). The first two lines in Figure 6.8(b)
show how a nonBI instruction is loaded into a register (*cInst*). Line three of Figure 6.8(b)
loads the last stored checksum and line four incrementally calculates the checksum and
line five stores it in *iChkSum*. Sixth line reset the flag *fBB*. Figure 6.8(c) depicts implanted micro-instructions for CFIs. Apart from incrementally check-summing the CFI
itself (lines one to five), the micro-instructions in the CFI will also perform encryption on
the checksum (line six) before comparing them against *eChkSum* (line eight). A DES56
Core, added as a functional unit is used for the encryption. A mismatch in the comparison
will result in a *SIGCKSM* singal as demonstrated by line ten of Figure 6.8(c). micro-
instruction routines as in Figure 6.8 will form the logic of the processor (*IMPRES*) that
performs monitoring at runtime.

Figure 6.9 depicts the interface of a DES56 Core and the algorithm used inside the core
for encryption/decryption. The core is capable of performing single DES encryption and
decryption functions in electronic codebook (ECB) mode. DES56 is a block cypher operating on 64-bit blocks. A 64-bit key is used, on which every eight bit (usually, parity bit)
is ignored, giving actual key size of 56-bits. Sixteen rounds of XORing with sub-keys,

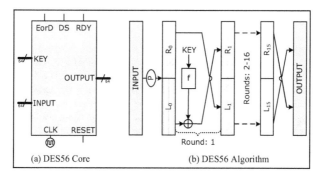

Figure 6.9: *DES56 Core Interface and Algorithm*

swapping the 32-bit blocks and table look-ups are performed before generating the final output. The reader is referred to [69] for further details on the DES56 algorithm and more.

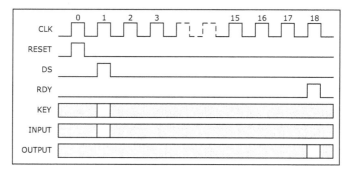

Figure 6.10: *DES56 Core Interface Timing*

The DES56 core is run in parallel with the CFIs to compute encrypted checksums as described in Figure 6.8. An encryption in the DES56 core for a 64-bit checksum takes 17 clock cycles as shown in the interface timing diagram in Figure 6.10. After the input block and the key are written to ports *INPUT* and *KEY* respectively and the DS (data strobe) signal is enabled, it takes 17 clock cycles to perform the encryption and the encrypted block (which is our encrypted checksum) becomes available on port *OUTPUT*. Furthermore, the *RDY* signal is enabled when the encrypted data becomes available. The clock period of DES56 core is small enough, so that we are able to use a clock divisor to generate a higher frequency clock for DES56 core and therefore perform the encryption in a single cycle of IMPRES processor. The clock period figures of the DES56 core and IMPRES are reported in Section 6.5.1.

The final task in the architectural design process is to generate the hardware in a hardware description language for simulation (behavioral model) and synthesis (gate level model). The same hardware monitoring is re-implemented in a cycle accurate instruction set simulator for performance evaluation and fault injection analysis.

6.3.3   Out-of-Order Processors and IMPRES

Our design process of IMPRES assumes an in-order processor model for runtime detection of software integrity violation. However, IMPRES is scalable to out-of-order processor models without further modifications to the IMPRES framework. Out-of-order execution is a paradigm used in most high-speed microprocessors in order to make use of clock cycles that would otherwise be wasted by stalls that occur in in-order processors, when the data needed to perform an operation is not available. Most of the modern commercial CPU designs include support for out-of-order execution.

Although the execution is performed out-of-order in an out-of-order processor, instruction fetch (IF), instruction decode (ID) and register write back (WB) stages are performed in-order. Even though, it is not possible to perform the check-summing and comparison operations during IF stage (since we trigger these operations based on the instructions, which are available during decode stage), it is possible to perform them during either ID or WB stages where the out-of-order processor behaves in-order and this will not affect IMPRES technique as proposed now.

6.3.4   Context Switching and IMPRES

Context switching is an activity performed by a processor for storing and restoring the state of the processor (the context) such that more than one processes can share a single processor resource. When the IMPRES processor is loaded with an operating system and more than one processes are executed, a register context switch could be performed between processes to enable IMPRES framework's activities. Special registers used to store intermediate values in IMPRES framework could be stored and restored between processes during context switching.

6.4   Experimental Setup and Evaluation

Although the method described in this chapter can be deployed in any type of embedded processor architecture, we have taken the PISA instruction set for our experimental implementation. The PISA instruction set is a simple MIPS like instruction set. IMPRES processor is developed by altering the rapid processor design process described in [211] for hardware synthesis (allowing a processor described in VHDL which is synthesizable). Applications from MiBench benchmark suite are used in the experiments. We have also performed fault injection analysis for single instruction corruption.

To evaluate IMPRES, applications from MiBench benchmark suite are taken and compiled with the GNU/GCC® cross compiler for PISA instruction set as described in Section 6.3.1. An automatic ASIP design tool, ASIP Meister [293] is used to generate the VHDL description of the target processor as described in Section 6.3.2. The outputs of ASIP Meister are the VHDL models for simulation for synthesis of IMPRES. The VHDL model is then altered to perform DES56 encryption during the executions of control flow

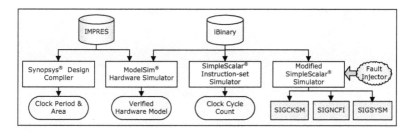

Figure 6.11: *Testing and Evaluation in IMPRES*

instructions. The synthesis model is used with Synopsys Design Compiler® (as shown in Figure 6.11) to obtain the area and clock period. Taiwan Semiconductor Manufacturing Company's (TSMC) 90nm core library is used for synthesis with typical conditions enabled. ModelSim hardware simulator is used with the instrumented binary (iBinary) to verify the correctness of the monitoring hardware.

We evaluated the clock cycle overhead of the proposed architecture using a cycle accurate instruction set simulator, SimpleScalar™ 3.0/PISA tool set [37]. The simulator is built around the existing cycle accurate simulator *sim-outorder*, in the tool set and used to calculate the clock cycle overheads. As depicted in Figure 6.11, the same simulator is modified to perform fault injection analysis and the results are tabulated and discussed in the following section. The micro-architectural parameters of SimpleScalar™ tool set were configured to model a typical embedded processor as designed by ASIP meister. The parameters used for the simulated processor are shown in Table 6.2.

Parameter	Value	Parameter	Value
Issue	in-order	Issue width	1
Fetch queue size	4	Commit width	1
L1 I-cache	16kB	L1 D-cache	16kB
L1 I-cache latency	1 cycle	L1 D-cache latency	1 cycle
Initial memory latency	18 cycles	Memory latency	2 cycles

Table 6.2: *Architectural Parameters for Simulation in IMPRES*

The parameters in Table 6.2 represent an in-order single issue processor (single issue width and single commit width) with instruction and data cache of single clock cycle latency. Level-1 instruction and data cache are enabled with the size of 16kB each. The clock cycle count from SimpleScalar™ tool set and the clock period from Synthesis Design Compiler® are used to calculate the total execution time of each of the application.

6.5   Results

In this section, we present memory and area overheads incurred by the proposed hybrid solution, as well as the impact of the technique on performance (total execution time). For the purpose of experiments, we have used the PISA instruction set as described in Section 6.4 and applications from the MiBench benchmark suite which represents typical workload for embedded processors.

6.5.1   Hardware Overhead

Table 6.3 shows the clock period and area overheads of IMPRES due to extra monitoring logic implementation. The clock period has increased a negligible 0.06% and the area increase is a paltry 5.03% that includes the encryption hardware, a DES56 core as discussed in Section 6.3.2. DES56 Core's clock period is 0.80ns and it runs in parallel to the regular processing and it will not effect IMPRES processor's clock period.

	Clock Period (ns)	Area (gates)
Ordinary H/W	16.84	227077
IMPRES H/W	16.85	229143
DES56 Core	0.80	9357
Total Area of IMPRES		238500
Overhead(%)	0.06%	5.03%

Table 6.3: *Clock Period and Area Comparison with and without IMPRES*

6.5.2   Performance Overhead

Table 6.4 reports the performance overhead incurred by our scheme for different applications (first column) from MiBench benchmark. In Table 6.4 columns 2 and 3 tabulate clock cycle comparison (in millions), columns 4 and 5 the execution time comparison (in seconds) and columns 6 and 7 the clock per instruction comparison. The columns with title *Ordinary* represents simulations of the ordinary processor hardware and *IMPRES* represents simulations of the IMPRES processor enabled.

Applications	Clock Cycle (10^6)		Execution Time (s)		Clock Per Instruction	
	Ordinary	IMPRES	Ordinary	IMPRES	Ordinary	IMPRES
adpcm.encode	120.6	131.2	2.03	2.21	1.62	1.47
adpcm.decode	90.0	97.6	1.51	1.64	1.57	1.40
blowfish.encrypt	79.2	79.6	1.33	1.34	1.36	1.27
blowfish.decrypt	80.4	80.8	1.35	1.36	1.36	1.28
crc32.checksum	57.6	57.6	0.97	0.97	1.36	1.20

Table 6.4: *Performance Comparisons with and without IMPRES*

As shown in Figure 6.11, the binary produced from software design process is used with an instruction set simulator (SimpleScalar® Tool Set) to compute the clock cycle overhead of our design.

$$Execution\ Time\ (s) = \frac{Clock\ Cycle\ \times Clock\ Period\ (ns)}{10^9} \quad (6.1)$$

As shown in Equation 6.1, the execution time of an application is computed by multiplying the clock cycle count reported by SimpleScalar™ and the clock period estimated by the hardware synthesis. For example, the execution time for application *adpcm.decode* on IMPRES processor is calculated by multiplying the clock period of the IMPRES processor (16.85ns) from Table 6.3 and clock cycle (97.6×10^6) count from Table 6.4 column two, giving 1.64s.

$$Clock\ Per\ Instruction = \frac{Clock\ Cycle}{Executed\ Code} \quad (6.2)$$

From Table 6.4, it is evident that the clock per instruction (defined by equation 6.2) reported by SimpleScalar™ for IMPRES is improved compared to the ordinary applications. That is, due to the instrumentation, the number of instructions executed per clock cycle in IMPRES have become higher compared to the applications without instrumentation. This improvement could be explained, as the instrumented instructions do not contain memory instruction that require more than one clock cycle to perform.

6.5.3   Memory Overhead

Application	Executed Code (10^6)		Code Size(# of lines)	
	Ordinary	IMPRES	Ordinary	IMPRES
adpcm.encode	74.4	89.8	402	460
adpcm.decode	57.4	69.8	397	452
blowfish.encrypt	58.3	62.6	2946	3085
blowfish.decrypt	59.0	63.2	2946	3085
crc32.checksum	42.5	48.0	527	607

Table 6.5: *Executed Code and Code Size Comparisons with and without IMPRES*

Table 6.5 compares the number of executed instructions and code size between IMPRES and ordinary processor for different applications (first column) from MiBench benchmark suite. In Table 6.5 columns 2 and 3 compare the number of executed instructions (in millions) and columns 4 and 5 compare the code size between the ordinary and IMPRES processors. The bigger the basic blocks in an application, the smaller the executed code and code size overheads, as each basic block carries an additional *chk* instruction. In applications used in the experiments, *blowfish* has bigger basic blocks compared to other applications. This explains the lower executed code and code size overheads of *blowfish* applications.

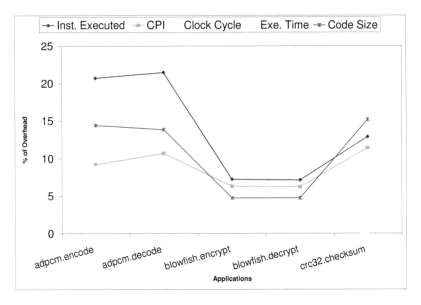

Figure 6.12: *Results Comparison for IMPRES*

As depicted in Figure 6.12, while the average increment in the number of instructions executed is 13.9%, the average clock cycle overhead is 3.66%. This could be explained by understanding that, all the extra instructions executed in IMPRES are non-memory instructions. Therefore, most of the clock cycle overhead of the extra instructions are hidden under the memory latency of the memory instructions of the original application. This claim is also addressed by the clock per instruction (CPI) improvements shown in Figure 6.12. Furthermore, as discussed earlier, all the overhead figures changes with the average sizes of the basic blocks in different applications. In Figure 6.12, the lines for clock cycle overheads and execution time overheads overlap each other as the clock period overhead is negligible.

6.6   Fault Injection Analysis

We performed the fault injection analysis by altering the SimpleScalar® tool set and adding fault injection scripts to the tool set as shown by the right side of Figure 6.11. Note that this fault injection analysis only checks for reliability and not security. However, since all the code injection attacks result in code integrity violations [184], it is feasible to use fault injection analysis as a measure for code injection attack detection.

The fault injection analysis is performed as follows:

1. after an application is securely loaded, a random address within the memory address boundaries is generated;

2. the instruction at this memory address is changed/corrupted with a randomly generated instruction; and

3. the application is executed and the output trace is analysed for fault activation and detection.

Applications	Not Act.	SIGSYSM	SIGCKSM	SIGNCFI	Total
adpcm.encode	2032	434	7396	138	10000
adpcm.decode	1595	414	7857	134	10000
blowfish.encrypt	4460	242	5233	65	10000
blowfish.decrypt	4314	267	5365	54	10000
crc32.checksum	4960	276	4650	114	10000

Table 6.6: *Fault Injection Results*

Table 6.6 shows the results of fault injection analysis. The three steps described above are performed 10000 times for each application and results are tabulated. The first column of Table 6.6 names the applications used for this analysis. Not all the faults injected are activated as some of them may fall into the non execution path of the program. These faults are called not activated (*Not Act.* in Table 6.6). Columns 3, 4 and 5 of Table 6.6 indicate the detected faults by different means. Column 3 (*System*) shows the faults detected by the instruction simulator itself (for example, an invalid opcode is detected by the system). Columns 4 and 5 in Table 6.6 are the number of faults detected by *SIGCKSM* and *SIGNCFI* respectively as described in Table 6.1.

Type	Activated At	Detected At	Δ	Γ(/bbsize)
T01	1	bbsize	bbsize-1	1
T02	1	1	0	1
T03	1	1	0	1
T04	1	1	0	1
T05	bbsize	bbsize	0	1
T06	bbsize	bbsize	0	1
T07	bbsize	bbsize+1	1	1
T08	bbsize	bbsize	0	1
T09	bbsize/2	bbsize	bbsize/2	bbsize-2
T10	bbsize/2	bbsize/2	0	bbsize-2
T11	bbsize/2	bbsize/2	0	bbsize-2
T12	bbsize/2	bbsize/2	0	bbsize-2

Table 6.7: *Error Detection Latency*

Table 6.7 tabulates the error detection latency and error probability due to bit bursts in a BB as modeled in Section 6.2.1). In Table 6.7, column one represents the types of code integrity violations (as modeled in Table 6.1 types T01-T12*); while column two contains the line in a BB at which the violation has occurred (Φ) , column three contains the line at which it is detected (Ψ) (if a particular violation is not linked to a particular line, an average position is assumed). Column four in Table 6.7 contains the error detection latency ($\Delta = \Phi - \Psi$, the number of clock cycles elapse between an error activation and detection) and column five shows the probability of occurrence of each type of violations (Γ), assuming a uniform distribution of bit bursts. *bbsize* in Table 6.7 indicates the last line of a BB or the number of instructions (the size) in a BB.

$$Average\ Error\ Detection\ Latency = \sum_{x=T01}^{T12} \Delta(x) \times \Gamma(x)/12 \qquad (6.3)$$

Assuming that the probabilities of all the code integrity violations modeled by T01-T12 in Table 6.1 are equal, the average error detection latency for a bit burst could be given by Equation 6.3. For example, for an average BB size of 8 instructions, the average error detection latency from Equation 6.3 is 1/3 instructions. Obviously, the error detection latency increases with the increase of basic block size.

It is worth noting that, the IMPRES framework is a code integrity violation detection scheme. When errors are detected by IMPRES, it is necessary to call an error recovery mechanism to recover from the errors. The fact that the error detection latency of IMPRES is very small as demonstrated above, makes it easier to implement an error recovery mechanism effectively. However, this is beyond the scope of this project.

6.7   Future Work

The basic block integrity violation model for IMPRES, as discussed under Section 6.2.2 has not been formally proven complete. We identify this as a future work. Further, implementing IMPRES on a superscalar processor has also been identified as a future work.

6.8   Summary of this chapter

In this chapter, we have presented a simple hardware-assisted runtime technique to detect code integrity violations. We have formulated a systematic model for basic block integrity violations and have shown that our scheme covers all the possible violations. Further, we have evaluated the fault coverage of our system via fault injection analysis and discussed the fault detection latency. We have elaborated a technique to design our solution using an ASIP design tool and software instrumentation. The hardware overhead and clock period change are evaluated and reported. The results show that these overheads are very small and negligible. Our study reveals that the proposed IMPRES framework is capable

---

*Types T13 and T14 in Table 6.1 are ignored here as they will not occur due to bit bursts and they are considered for code injection attack.

of handling code injection attacks and transient bitflips via detecting code integrity violations with minimal overheads. We conclude that our technique is useful in answering the increasing security and reliability concerns in embedded systems.

Chapter 7

A Hybrid Hardware/Software Technique for Preemptive Control Flow Checking in Embedded Processors

... 'I wonder if I've been changed in the night? Let me think. Was I the same when I got up this morning? I almost think I can remember feeling a little different. But if I'm not the same, the next question is 'Who in the world am I?' Ah, that's the great puzzle!'...

— Lewis Carroll, *Alice's Adventures in Wonderland*

Numerous methods have been described in literature to improve reliability of processors by the use of control flow checking (CFC). High performance and code size penalties cripple the proposed software only approaches, while hardware approaches are not scalable and are thus rarely implemented in real embedded systems. In this project, we show that by including control flow checking as an issue to be considered when designing as embedded processor, we are able to reduce overheads considerably and still provide a scalable solution to this problem. The technique used in this project includes architectural improvements to the processor and binary rewriting of the application. Architectural refinement incorporates additional instructions to the instruction set architecture, while the binary rewriting utilizes these additional instructions into the program flow. Applications from an embedded systems benchmark suite have been used to test and evaluate the system. Our approach increased code size by only 5.55-13.5% and reduced performance by just 0.54-2.83% for eight different industry standard benchmarks. The additional hardware overhead due to the additional instruction in the design is just 2.70%. In contrast, the state of the art software-only approach required 50-150% additional code, and reduced performance by 53.5-99.5% when monitoring was inserted. Fault injection analysis demonstrates that our solution is capable of capturing and recovering from all the injected control flow errors (CFEs).

131

7.1   About this Chapter

7.1.1   Objectives

This chapter has the following primary goals:

1. it is possible to craft a hardware assisted and still scalable control flow checking technique;

2. considering reliability as a design time issue will reduce the cost of that technique; and

3. showing the overheads and fault coverage of the proposed control flow error detection technique.

7.1.2   Outline

This project presents a hardware technique to detect control flow errors (CFEs) at the granularity of micro-instructions. We use a hybrid hardware software technique to deal with this problem and show that we are able to reduce the overheads compared to the state of the art techniques. Our scheme uses micro-instruction routines to perform runtime hardware monitoring for detecting CFEs and binary rewriting for the software instrumentation. Micro-instructions are instructions which control data flow, and instruction-execution sequencing in a processor at a more fundamental level than at the level of machine instructions. The software specialization performed in our scheme is minimal compared to the software instrumentations in the software only approaches, because the specialization is used only as an interface between check points and micro-instruction routines. Our checking architecture could be deployed in any embedded processor as a design parameter to observe its control flow at runtime and trigger a flag when any unexpected control flow pattern is detected.

In this project, we address fault tolerance by focusing on the specific problem of ensuring correct execution of expected control flow of a program. We have evaluated memory, area and clock period (reduction in frequency) overheads associated with the proposed architecture using applications from an embedded systems benchmark suite called MiBench [115]. Hardware synthesis and simulations are performed by commercial design tools. Results demonstrate that our proposed solution has significant reduction in the overheads compared to other techniques in the literature. The approach discussed in this project reduces the overheads by trading a few more transistors to achieve performance gain and memory overhead. Our binary rewriting approach solves the deficiency caused by uninsturmented library functions as we rewrite the whole binary, including the library functions used in the particular application.

Generally, hardware assisted schemes are not scalable as they perform monitoring by observing the memory access patterns using watchdog processors. These schemes have to be tailor made for each application. Our scheme is scalable like a software-only approach due to the software specialization we perform, however, unlike software-only approaches, our method incurs very little code size and performance overheads with superior error coverage.

7.1.3 Contributions and Limitations

The following are our contributions reported in this project. We have proposed a method that,

1. is scalable (unlike most of the hardware assisted techniques) for embedding CFE detection at the granularity of micro-instructions;

2. not only detects random bit flips but also bit bursts occurring in instruction memory, data bus & some of the registers;

3. is scalable like software only approaches, but incurs very little code size and performance overheads;

4. shares most of the monitoring hardware with the regular processing and therefore requires very little additional hardware; and

5. uses binary rewriting to insert check instructions and therefore requires no compiler support and could be applied equally to newly compiled applications and legacy code.

The limitations of our proposed approach and the solutions to cope with these limitations are given bellow:

1. Our scheme will not capture CFEs caused by a corrupted non CFI turning into a Control Flow Instruction (CFI). However, our claim of making CFC a design parameter, solves this limitation. By designing the CFIs such that the hamming distances between the opcodes of non CFIs and CFIs higher, it is unlikely that a non CFI will turn into a CFI.

2. Our solution assumes full control over the hardware design process. Our decision to dictate CFC as a design parameter supports our assumption. Thus this approach is ideal for future processors.

3. We assume an in-order processor model in our design. However, it is possible to use our solution in an out-of-order processor by performing the calculations and checks on the in-order part of the processing pipe (i.e the decoding or the write back stage).

4. Our technique is not applicable to off-the-shelf processors, for which software-only solutions are the only approach to follow (with associated overheads).

5. Binary rewriting will only insert check instructions into the user code and statically bound library functions. However it will not insert checks into dynamically shared libraries. However, it is possible to instrument the dynamically shared libraries, by

instrumenting the operating system kernel using the same software instrumentation process described in this project.

6. Our scheme will not detect CFEs caused by corruptions in the internal registers including pipeline registers. It is possible to detect some corruptions by using parity enabled internal registers as proposed in [97].

7.1.4   Rest of this chapter

The remainder of this chapter is organized as follows. Section 7.2 presents the proposed error checking architecture. Section 7.3 describes a systematic methodology to design the proposed solution for a given architecture. The implementation and evaluation are presented in Section 7.4. Results are presented in Section 7.5 and a summary of this chapter is presented in Section 7.6.

7.2   Control Flow Checking Architecture

In this section, we provide an overview of our hybrid hardware-software control flow checking architecture. We then describe how the architecture works at runtime to detect control flow errors.

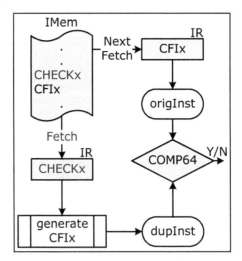

Figure 7.1: *Control Flow Checking Architecture*

Figure 7.1 depicts the conceptual flow diagram of the proposed checking architecture. For ease of illustration, we depict only the hardware units related to the checking architecture with respect to the whole datapath of a processor. *IMem* in Figure 7.1 represents the instruction memory segment of the processor. Each CFI of a given application (indicated by CFIx in Figure 7.1) is preceded by a check instruction (indicated by CHECKx in Figure 7.1). This instrumentation is performed by a software component using binary rewriting. Check instruction passes information needed to reconstruct a duplicate copy of the CFI, which follows the check instruction. More details on software binary rewriting is discussed in Section 7.3.

At runtime, when a check instruction (CHECKx) is executed, it handles three distinctive tasks. They are:

1. assembling a duplicate copy of the CFI of concern from the information that the check instruction carries and write it into a special register (*dupInst* in Figure 7.1);

2. copies the following CFI that is fetched during the next fetch cycle into another special register (indicated by *origInst* in Figure 7.1);

3. comparing the values in *dupInst* and *origInst* using a comparator and sending the result to a flag. Note that for register indirect addressing, the values of the registers also have to be duplicated and compared.

For CFIs with register indirect addressing, it is essential to verify the contents of the registers apart from the binaries of the CFIs themselves. The rudimentary solution is to have a shadow register file and make each register writeback to write to both the real and the shadow register files. When a register is used in a CFI, the duplicate CFI will not only perform a comparison between the binaries of the instructions, but also perform comparisons between the real and shadow registers used by the CFIs. Performing writebacks to shadow registers for each instruction will involve huge amount of unwanted switching activity. This unwanted switching activity could be reduced by performing shadow register writebacks at only necessary points (points where registers used in CFIs are updated) in an application program. Register updates related to CFIs could be identified by using the *use-def chains* [*] (register definitions) of a particular application.

Micro-instructions of the check instruction are formed such that the check instruction will perform the tasks mentioned above in the given order. Our method's preemptive error detection property is demonstrated in this approach, since a fault in a CFI is detected before the erroneous instruction itself is executed.

The flag mentioned in the third task above is checked to see if there is a mismatch in the CFIs. If the flag is set, then it indicates the detection of a control flow error. This flag could be used not only to detect errors in the CFIs but also to recover from CFEs assuming that the CFIs are error free. Even though the errors introduced in the check instructions will affect the recovery mechanism, they will still guide detecting errors introduced within the instruction memory (a corruption in the check instruction should not be considered a false alarm in terms of error detection, because some underlying cause for the corruption exists and this may well affect a legitimate instruction in the future).

---

[*]Data structures which model the relationship between the definitions of variables, and their uses in a sequence of assignments

7.3   Design Flow

In this section, an overview of the proposed design flow for the control flow checking architecture is provided. First, the design of a software interface that allows the applications to interact with the architectural enhancement is described, and then the design of the architectural enhancement itself is discussed.

7.3.1   Software Design

Figure 7.2 describes the implementation details of the interface between an application program and the fault checking hardware. It is worth noting that the check instructions inserted at right places in an application serves as the interface between software and the fault checking hardware. The software instrumentation process proposed by the use of binary rewriting technique is described in [284]. This is a code specialization technique that performs an X-to-X binary translation. In [284] the authors propose binary rewriting in as compiler tool-chain extension, called Diablo for code compaction, an unrelated application of binary rewriting.

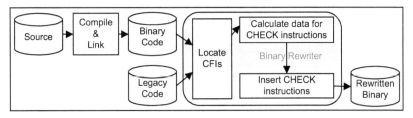

Figure 7.2: *Software Design flow for the control flow checking architecture*

The tasks assigned to be performed by the binary rewriter are:

1. locating all the CFIs and determining their types based on opcodes;

2. extracting information necessary to re-construct the same CFI at runtime;

3. forming the check instructions using the data from task 2 above; and

4. finally, placing the check instructions into the application binary using the binary rewriting technique.

Inserting signatures at not uniform intervals into an application binary and using these signatures at runtime to perform hardware checks has never been possible without compiler support [183]. Since CFIs (and therefore inserted check instructions) are not going to be in regular intervals in an application program, we propose a binary rewriting technique [284] to perform software instrumentation.

Figure 7.3 depicts the positions where compile time instrumentation and binary rewriting endure in a programming tool chain. Compile time instrumentation will only be capable

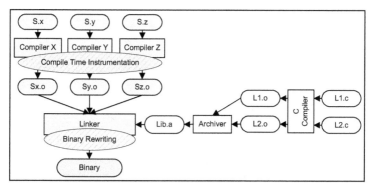

Figure 7.3: *Compile Time Instrumentation versus Binary Rewriting*

of asserting checks within the user code and not functions included from library archives. On the other hand, using binary rewriting it is possible to assert checks not only in user functions but also library functions. We are assuming that all code constituting a program is available at link time, or in other words, that the programs are statically bound. This assumption is not valid for dynamic languages such as Java, where, because of reflection, the complete program may not be available. Dynamic shared libraries, of which the code is only available when a program is loaded into memory also violate this assumption. However, performing the instrumentation with binary rewriting will not only insert check instructions for library functions, but also enables instrumentation of legacy binary code. Furthermore, a binary rewriter will not require compiler modifications and therefore the same compilers could still be used without any changes.

The opcodes and instruction format of check instructions are made offline, and is only performed once for a particular ISA at design time. When this decision is made, a binary rewriter as described above could be implemented and then the whole software design process is made automatic.

7.3.2 Architectural Design

Without losing the generality of our technique, we use an automatic processor design tool to implement the reliable processor in hardware. This automatic design tool is used to design Application Specific Instruction-set Processors (ASIPs)[91], custom designed for applications or application domains.

Figure 7.4 describes the hardware design process of the hybrid CFC architecture of a processor. Check instructions are also made part of the instruction set of the ISA. In the design tool, the functional units required to implement the processor will be chosen from a resource pool. Using the information of the ISAs for check instructions, micro-instruction routines are formed, and are included into the check instruction implementation. These micro-instruction routines will form the logic of the processor that will perform the CFC at runtime. The final task in the architectural design process is to generate the hardware model of the processor in a hardware description language for simulation (behavioral) and synthesis (gate level) (indicated by *CFC Processor Model* in Figure 7.4).

Figure 7.5 depicts the micro-instruction routine of the check instruction that will be inserted along with a *j* *<targ>* instruction. *j* *<targ>* is a typical 'jump to an immediate' in a RISC type processor. In this particular case, the opcode of the *j* *<targ>* is *0x0001* and the target (*targ*) is a 26bit immediate value. The instruction fetch stage of the micro-instruction routine (IF) loads the current instruction (currentInst) from the instruction memory via instruction memory access unit (IMAU) and writes it to the instruction register (InstReg). The address from where the current instruction is fetched is obtained from the program counter (CPC), which is incremented at the end of an instruction fetch stage.

Instruction Decode (ID) stage of the check instruction as shown in Figure 7.5 generates a duplicate copy of the following CFI (dupInst) from the pre-defined values and the target value (targ), which is an argument to the current instruction. At the execution (EX) stage of the pipeline, the original CFI is read from the instruction register (note that at the EX stage of the pipeline of the current instruction, the *InstReg* contains the binary of the instruction fetched next) and compares the original instruction (origInst) with the generated duplicate copy (dupInst). A mismatch is stored in a flag and will be used to detect and/or recover from a CFE. The check instruction described here uses a comparator *COMP64*, which is a functional unit that is shared among the other instructions.

7.4   Implementation and Evaluation

Although the method described in this project for control flow checking can be deployed in any type of embedded processor architecture, we have taken the PISA instruction set for our experimental implementation. The PISA instruction set is a simple MIPS like instruction set. Control flow checking in the processor is enabled by altering the rapid processor design process described in [211] for hardware synthesis (allowing a processor described in VHDL which is synthesizable).

To evaluate our approach, applications from MiBench benchmark suite are taken and compiled with the GNU/GCC® cross-compiler for PISA instruction set. As mentioned previously, an automatic processor design tool, called ASIP Meister [293] is used to generate the VHDL description of the target processor as described in section 7.3.2. The output of the ASIP Meister are the VHDL models of the processor for simulation and synthesis. As shown in Figure 7.6, the rewritten binary produced from the software de-

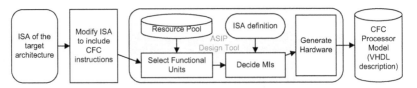

Figure 7.4: *Hardware Design flow for the control flow checking architecture*

```
wire [63:0]    dupInst;

wire [31:0]    currentPC;
wire [63:0]    currentInst;

currentPC      = CPC.read()                              IF
currentInst    = IMAU.read(currentPC);
null           = InstReg.write(currentInst);
null           = CPC.inc();

wire [15:0]    opcode;
wire [15:0]    zeros16;
wire [5:0]     zeros06;

opcode         = "0000000000000001";                     ID
zeros16        = "0000000000000000";
zeros06        = "000000";
dupInst        = <zeros16,opcode,zeros06,targ>;

wire [63:0]    origInst;

origInst       = InstReg.read();                         EX
flagError      = COMP64.cmp(dupInst, origInst);
```

Figure 7.5: *Micro-instruction Routine of a Check Instruction* (chk31 <targ>) *for a CFI (* j <targ>)

sign and the processor simulation model generated from hardware design are used in ModelSim® hardware simulator to verify the functional correctness of our design. The gate level VHDL model from ASIP Meister is used with Synopsys Design Compiler® to obtain the clock period and area overheads.

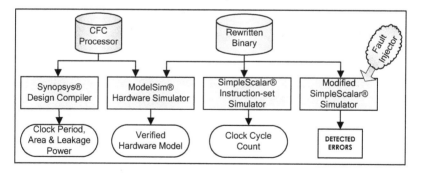

Figure 7.6: *Testing and Evaluation of the CFC*

We evaluated the clock cycle overhead of the proposed architecture using a cycle accurate instruction set simulator, SimpleScalar™ 3.0/PISA tool set [37]. The simulator is built around the existing cycle accurate simulator *sim-outorder*, in the tool set and used to calculate the clock cycle overheads. As depicted in Figure 7.6, the same simulator is modified to perform fault injection analysis and the results are tabulated in the next Section. The micro-architectural parameters of SimpleScalar™ were configured to model a typical embedded processor as designed by ASIP meister. The parameters used for the simulated processor are shown in Table 7.1.

The clock cycle counts from SimpleScalar™ and the clock period from Synopsys Design Compiler® are used to calculate the total execution times of all the applications.

7.5   Experimental Results

In this section, we present memory and area overheads incurred by the proposed hybrid solution, as well as the impact of the technique on performance (total execution time). Further, we give results from the fault injection analysis performed on our processor model. For the purpose of experiments, we have used the PISA instruction set as described in Section 7.4 and applications from the MiBench benchmark suite which represents a typical workload for embedded processors.

7.5.1   Hardware Overhead

Table 7.2 tabulates the area and clock period overheads due to the enhancement in the hardware. The overheads here represent the extra logic in our design compared to the design without CFC implemented. Synopsys Design Compiler® with Taiwan Semiconductor Manufacturing Company's (TSMC) 90nm core library with typical conditions enabled is used for synthesis. The percentage of overheads of area is 2.70% and clock period is 0.53% compared to a processor which is designed without CFC capabilities.

7.5.2   Performance Overhead

Table 7.3 reports the performance overhead incurred by our scheme for different applications (first column) from MiBench benchmark. In Table 7.3, columns 2-4 tabulate clock cycle comparisons (in millions) and percentage of overheads and columns 5-7 tabulate execution time comparison (in seconds) and percentage of overheads. The columns with title *NoCHK* represents simulations of the processor without control flow checking enabled and *CHK* represents simulations of the processor with control flow checking enabled. Clock cycle overhead ranges from 0.00% to 2.28% with an average of 1.19% and execution time overhead ranges from 0.54% to 2.83% with an average of 1.73%. In con-

Parameter	Value	Parameter	Value
Issue	in-order	Issue width	1
Fetch queue size	4	Commit width	1
L1 I-cache	16kB	L1 D-cache	16kB
L1 I-cache latency	1 cycle	L1 D-cache latency	1 cycle
Initial memory latency	18 cycles	Memory latency	2 cycles

Table 7.1: *Architectural Parameters for Simulation of the CFC*

Parameters	Without Checks	With Checks	% overhead
Area (cells)	228489	234669	2.70
Clock Period (ns)	16.85	16.94	0.53

Table 7.2: *Hardware Overhead for the Control Flow Checking*

trast, the performance (execution time) overhead in software-only PECOS technique was in the range of 53.5-99.5% (though their method used existing processors).

Benchmarks	Clock Cycle/10^6			Execution Time/s		
	NoCHK	CHK	%	NoCHK	CHK	%
adpcm.decode	121.6	123.0	1.15	2.05	2.08	1.46
adpcm.encode	89.96	90.64	0.76	1.52	1.54	1.30
blowf.encrypt	79.21	79.90	0.87	1.33	1.35	1.41
blowf.decrypt	80.44	81.25	0.89	1.36	1.37	1.42
crc32.checksum	57.62	57.63	0.00	0.97	0.98	0.54
jpeg.compress	16.41	16.66	1.54	0.28	0.28	2.09
jpeg.decompress	10.79	11.01	2.02	0.18	0.19	2.57
jpeg.transcoding	8.96	9.17	2.28	0.15	0.16	2.83

Table 7.3: *Performance Comparisons with and without CFC*

$$Execution\ Time\ (s) = (Clock\ Cycle \times Clock\ Period\ (ns))/10^9 \qquad (7.1)$$

As shown in Equation 7.1, the execution time of an application is computed by multiplying the clock cycle count reported by SimpleScalar[TM] and the clock period estimated by our hardware synthesis. For example, the execution time for application *adpcm.decode* with the check instructions inserted is calculated by multiplying the clock period of the processor with checks (16.94ns) from Table 7.2 and clock cycle count (123.0×10^6) from Table 7.3, giving 2.08s.

7.5.3 Codesize Overhead

Table 7.4 reports the static and executed code size overheads resulting from our software instrumentation for eight different applications (first column). In Table 7.4 columns 2-4 tabulate executed instruction count comparisons (number of lines in millions) and percentage of overheads, and columns 5-7 tabulate static instruction count (number of lines) comparisons and percentage of overheads. The columns with title *NoCHK* represents simulations of the processor without control flow checking enabled and *CHK* represents simulations of the processor with the control flow checking enabled. Executed instruction count overhead ranges from 6.41% to 20.2% with an average of 13.5% and static instruction count (code size) overhead ranges from 5.55% to 13.3% with an average of 10.1%. The code size overhead of software only PECOS technique was in the range of 50-150%.

Benchmarks	Executed Inst./10^6			Static Inst.		
	NoCHK	CHK	%	NoCHK	CHK	%
adpcm.decode	76.55	92.98	17.7	623	694	11.4
adpcm.encode	58.91	72.60	18.9	618	688	11.3
blowf.encrypt	58.43	64.44	9.33	6463	6822	5.55
blowf.decrypt	59.04	65.05	9.24	6463	6822	5.55
crc32.checksum	42.51	50.72	16.2	527	563	6.83
jpeg.compress	11.62	12.88	9.83	58650	66581	13.5
jpeg.decompress	7.45	7.96	6.41	56577	64048	13.2
jpeg.transcoding	5.91	7.41	20.2	52605	59593	13.3

Table 7.4: *Codesize Comparison with and without CFC*

From Tables 7.3 and 7.4, while the average increase in number of instructions executed is 13.5%, the average clock cycle overhead is 1.19%. This could be explained by understanding that all extra instructions executed during CFC are non-memory instructions. Therefore, most of the clock cycle overhead of the extra instructions are hidden under the memory latency of the memory instructions from the original application.

7.5.4   Fault Injection Analysis

Table 7.5 tabulates the results of the fault injection analysis performed on our control flow detection architecture. Random memory errors are generated by performing bit flips in the instruction memory. Random memory errors represent a wide range of transient errors in hardware and some errors in software. For each application, ten thousand (10,000) errors were inserted one at a time and the runtime program behavior was observed to identify whether the error was captured by our technique. With the current design, all the errors injected into the CFIs were successfully detected and could be recovered without further effort by our system. In contrast, the software only PECOS method detected 87% of all the CFEs. Furthermore, errors injected into the check instructions in the instruction memory will also be detected by our scheme, although we need an additional recovery mechanism to recover from these errors.

In Table 7.5, column one contains the names of the applications used for fault injection analysis and columns 2, 3 and 4 contain the total number of injected errors, the number of activated errors and the number of activated CFEs respectively. If a faulty instruction is executed during a particular execution of the application then that error is categorized as activated and otherwise it is not. Column 6 of Table 7.5 represents the number of CFEs captured (note that, this is equal to the number of CFEs present in the system). All the captured CFEs are recoverable by our scheme without further effort by assuming the check instructions in the memory are error free. Column 5 of Table 7.5 contains the total number of errors detected by our scheme. Although the faults injected in non CFIs do not represent CFEs, they do represent some underlying cause for the corruption in the system and this may well be considered detected faults. By implementing a recovery mechanism and using our detection scheme, it is possible to recover from all the detected errors.

Benchmarks	Injected Errors			Detected Errors	
	Total	Activated	CFEs	Total	CFEs
adpcm.decode	10000	5723	769	1504	769
adpcm.encode	10000	4635	607	1235	607
blowf.encrypt	10000	6242	911	1880	911
blowf.decrypt	10000	6283	949	1910	949
crc32.checksum	10000	6264	969	1917	969
jpeg.compress	10000	2947	417	842	417
jpeg.decompress	10000	2901	428	851	428
jpeg.transcoding	10000	4176	710	1421	710

Table 7.5: *Fault Injection Analysis for CFC*

This fault injection analysis is only capable of testing our design for instruction memory bit flips and bursts detection. Further analysis could be performed to test the capability of the detection of errors in registers used by CFIs. However, all the injected errors in the register file will be captured by our scheme given that the CFIs which use the errornous registers are checked for errors. Therefore, the error coverage of the register file is equal to the error coverage of the CFIs and the registers used in the CFIs.

7.6 Summary of this chapter

In this project, we have presented a design time hardware software technique to detect CFEs caused by bit flips and bursts at runtime before an erroneous CFI is executed. We have formulated a formal method to accommodate this technique within an automatic embedded processor design flow. Our evaluation studies reveal that the solution we have proposed is capable of handling CFEs with as little as 2.70% of area and 1.73% of performance overheads. These overheads are minimal compared to the software solution that deals with the same problem. We conclude that by asserting CFC as a design requirement of an embedded processor, it is feasible to reduce the overhead of CFC to as minimal as possible.

Fault injection analysis demonstrates that our solution is capable of capturing and recovering (without any further effort) from all the injected CFEs in the instruction memory. Furthermore, our scheme is capable of detecting errors injected into the check instructions. Our error detection scheme is preemptive and is capable of correcting CFEs (assuming the error is in the original CFI) without any additional overheads. We believe that the technique described in this project could be used with any embedded processors for detecting CFEs efficiently.

Chapter 8

Hardware Assisted Preemptive Control Flow Checking for Embedded Processors to improve Reliability

... 'it would be so nice if something made sense for a change.'

— Lewis Carroll, *Alice's Adventures in Wonderland*

The work presented in this chapter is an extension to the one presented in hybrid hardware/software technique for preemptive control flow checking (CFC) in embedded processors in the previous chapter (Chapter 7). Therefore, their exists a small overlap between these two chapters in describing the architecture.

Reliability in embedded processors can be improved by control flow checking and such checking can be conducted using software or hardware. All the proposed software-only approaches suffer from significant code size penalties, resulting in poor performance. Most of the proposed hardware-assisted approaches are not scalable and therefore cannot be implemented in real embedded systems. This chapter presents a scalable, cost effective and novel fault detection technique to ensure proper control flow of a program. This technique includes architectural changes to the processor and software modifications. Architectural refinement incorporates additional instructions, while the software transformation utilizes these additional instructions into the program flow. Applications from an embedded systems benchmark suite are used for testing and evaluation. The overheads are compared with the state of the art approach that performs the same error coverage using software-only techniques. Our method has greatly reduced overheads compared to the state of the art. Our approach increased code size by between 3.85-11.2% and reduced performance by just 0.24-1.47% for eight different industry standard benchmarks. The additional hardware (gates) overhead in this approach was just 3.59%. In contrast, the state of the art software-only approach required 50-150% additional code, and reduced performance by 53.5-99.5% when monitoring was inserted.

8.1 About this Chapter

8.1.1 Objectives

This chapter has the following primary goals:

1. providing a different (to that presented in Chapter 8) code duplication technique to detect control flow errors (CFE); and

2. providing additional hardware architectural support to detect control flow errors due to register-file and program counter corruptions.

8.1.2 Outline

This chapter presents a hardware software technique to detect control flow errors at the granularity of micro-instructions (MI). By using such a technique to deal with this problem and we are able to reduce the overheads to a considerable minimum. micro-instructions are instructions which control data flow, and instruction-execution sequencing, in a processor at a more fundamental level than the level of machine instructions. The software instrumentation performed in our scheme is minimal compared to software only approaches, because it is used only as an interface between check points and micro-instruction routines. Our checking architecture could be deployed in any embedded processor on which we have the design control to observe its control flow at runtime and trigger a flag when any unexpected control flow pattern is detected.

Parity checking as implemented in many other well known solutions [97], provides good protection against single bit errors when the probability of errors are independent. However, in many circumstances, errors come in groups, which we call bit bursts. Parity checking provides very limited protection against bit bursts. The technique proposed in this project detects control flow error caused by not only independent bit flips, but also bit bursts.

Figure 8.1 compares the software instrumentations performed during the control flow error detection proposed in the previous chapter (Chapter 8) and the extension proposed in this chapter. Figures 8.1(a).(i) and (b).(i) depict code fragments of a typical RISC processor. Figures 8.1(a).(ii) and (b).(ii) depict the same code fragment after the software instrumentation (SI) for detecting bit flips in instruction memory. The technique proposed in Chapter 8 used *check* instructions to perform the duplication and the current extension uses the same control flow instructions (CFIs) and some hardware modifications to perform the duplication (details are given later in this chapter).

For example, in Figure 8.1(a) a control flow instruction (CFI), *bgez $2,$L6* will be instrumented with *chk1 $2,$L6*. The opcode of instruction *chk1* itself will point out the type of control flow instruction that is going to follow (in this example, a conditional branch *bgez*) and the arguments to the check instruction will carry the information necessary to generate a duplicate copy of *bgez $2,$L6*. Another example from Figure 8.1 is *j $L7*

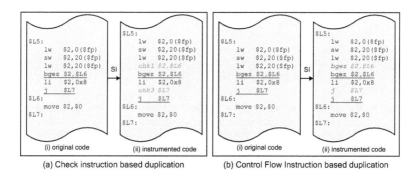

Figure 8.1: *Detecting Bit Flips in Instruction Memory*

and *chk3 $L7*. We use micro-instruction routines to form *chk1* and *chk3*, such that the micro-instructions will form duplicate copies of the following control flow instructions and compare them with the originals at runtime. In Figure 8.1(b), the duplicate copy of the same instruction is inserted just before the original control flow instruction. At runtime, the duplicate copy of the control flow instruction will perform a comparison against the original fetched instruction and will send an error if there is a mismatch. The error signal is sent before the original control flow instruction is executed, thus making the error detection preemptive.

The technique presented here is preemptive control flow error detection, which prevents process crashes (a crash of the entire process incurs a higher recovery overhead due to the overhead of process creation) and error propagation (latency of propagated errors is at least several hundreds of instruction cycles [134] and errors which are not detected early may cause severe problems like check point corruption which complicates error recovery).

8.1.3   Contributions

Our contributions in this work are:

1. detection and correction of bit bursts that causes control flow error unlike parity based hardware assisted techniques;

2. preemptive and correctable control flow error detection as opposed to signature based hardware assisted mechanisms;

3. a scalable mechanism as we use software instrumentation as the interface for CFC;

4. a methodology for embedding control flow error monitoring at the granularity of micro-instructions;

5. a technique that requires very little code size overhead and performance overheads compared to the software-only approaches; and

6. a method to share most of the monitoring hardware and therefore requires very little additional hardware.

8.1.4   Rest of this chapter

The remainder of this chapter is organized as follows. Section 8.2 presents the proposed error checking architecture. Section 8.3 describes a systematic methodology to design the proposed solution for a given architecture. Implementation and evaluation are presented in Section 8.4. Results are presented in Section 8.5 and a summary of this chapter in Section 8.7.

8.2   Error Checking Architecture

In this section, we provide an overview of our hardware architecture for CFC. The hardware modifications that are performed to enable CFC on a pipelined RISC architecture are:

1. enhancements to the controller to treat control flow instructions to handle instruction memory bit flips;

2. an addition of a shadow register file and the related logic to handle control flow errors caused by indirect control flow instructions; and

3. an inclusion of a shadow PC and the accompanying logic to handle program counter corruption that causes control flow errors.

In this section, we also describe how the architecture works at runtime to detect control flow errors.

8.2.1   Instruction Memory Bit Flips Detection

Figure 8.2 depicts the conceptual flow diagram of the proposed instruction memory bit flips detection and correction mechanism. For ease of illustration, we only depict the hardware units related to the checking architecture with respect to the whole architecture of an embedded processor. *IMem* in Figure 8.2 represents the instruction memory segment of the processor. Each control flow instruction of a given application (*CFIo*) is preceded by a duplicate copy of the same instruction (*CFId*). This instrumentation is performed by a software component at compile time. The pipeline stages shown in the upper part of Figure 8.2 belongs to *CFId* and the lower part belongs to *CFIo*. Each fetch writes the binary of the instruction fetched into the instruction register (IR) of the processor. Whether a fetched control flow instruction is either original or duplicate is decided by the processor by checking a special single bit flag that flips back and forth for each control flow instruction. A duplicate control flow instruction, *CFId* in its execution stage (EXE), compares its own binary against the one fetched next to it and signals an error when there is a mismatch.

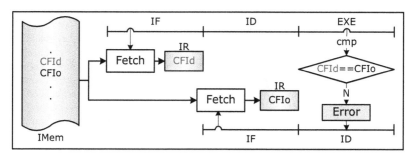

Figure 8.2: *Control Flow Checking Architecture*

Micro-instructions of the control flow instruction are formed such that the control flow instruction will perform the tasks as described above in the same order. The outcome from the comparison between *CFIo* and *CFId* could be used to either detect the error and stop the application or correct the control flow instruction by assuming an error free duplicate control flow instruction. Our method's preemptive error detection property is demonstrated in this approach, since a fault in a control flow instruction is detected before the erroneous instruction itself is executed.

Figure 8.3 depicts how duplicate and original control flow instructions are distinguished by our processor when an instrumented application is executed. A single bit *flag register* *(F)* is added (or the current flag register could be extended) to the architecture of the processor which will be reset at the beginning. When control flow instructions including duplicates are encountered the flag will be flipped between zero and one. If the flag is set to one when a control flow instruction is fetched, then the fetched control flow instruction is deemed to be an original (*CFIo*) and otherwise it is deemed as a duplicate (*CFId*).

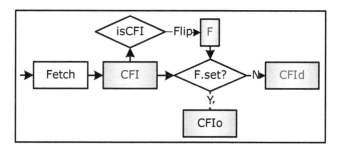

Figure 8.3: *Separating Original and Duplicate CFIs*

The same technique described in this section will also detect control flow errors caused by errornous transmissions between different memory hierarchies and the transient control flow errors in the instruction memory data bus.

8.2.2 Shadow Register File

For control flow instructions with register indirect addressing, it is essential to verify the contents of the registers apart from the binaries of the control flow instructions themselves. The rudimentary solution is to have a shadow register file and make each register writeback to write to both the real and the shadow register files. When a register is used in a control flow instruction, the duplicate control flow instruction will not only perform a comparison between the binaries of the instructions, but also perform comparisons between the real and shadow registers used by the control flow instructions. Performing writebacks to shadow registers for each instruction will involve huge amount of unwanted switching activity. This could be reduced by performing shadow register writebacks at only necessary points in an application program. This could be achieved by using the *use-def chains* * (register definitions) of a particular application, which is already present at compile time in all the optimizing compilers.

8.2.3 Shadow Program Counter

Control flow errors may also occur due to bit flips or a burst in the program counter (PC). We propose a shadow PC to overcome this. A shadow PC is included in the hardware and will be loaded and incremented synchronously with the real PC. When a PC read operation is performed, a copy of the PC value will also be read from the shadow PC and a comparison is performed between them. A mismatch will result in program abortion, or continuation with the assumption that the shadow PC is not corrupted. This scheme will give a better error coverage than a processor without a shadow PC.

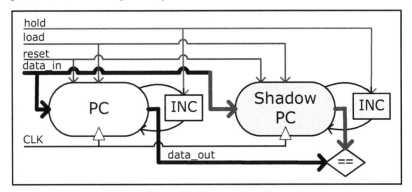

Figure 8.4: *Shadow Program Counter*

Figure 8.4 depicts the proposed architecture for a shadow PC. The input signals to the real PC are extended to the shadow PC and output from the real PC is compared against the output from the shadow PC to detect bit flips or a burst in the real PC.

---

*Data structures which model the relationship between the definitions of variables, and their uses in a sequence of assignments

8.3   Design Flow

In this section, an overview of the proposed design flow for the checking architecture is provided. First, the design of a software interface that allows the applications to interact with the architectural enhancement is described, and then the design of the architectural enhancement itself is discussed.

8.3.1   Software Design

Figure 8.5(a) describes the implementation details of the interface between an application program and the fault checking hardware. It is worth noting that the duplicate control flow instructions inserted at the right places in an application serves as the interface between software and fault checking hardware. In the software instrumentation process, the source code of an application is compiled by the front end of a compiler and the assembly code for the target ISA is produced. Then a software parser is used to instrument the assembly code. control flow instructions are located and duplicate copies of the control flow instructions are inserted into the application.

The software instrumentation described above is in the instrumentation process for control flow instructions with constant offsets. For control flow instructions with register indirect addressing, the register source will be duplicated by means of shadow registers at the time of register definition and used by the duplicate control flow instructions for comparison. As described in Section 8.2.2 this could be achieved by two different means and they are: [i] by enabling *shadow register writeback* whenever a register writeback is performed; or [ii] by generating and using special instructions to perform register writeback only in places of the application where a register writeback related to control flow instruction is performed. The former will incur additional unwanted switching activity and the later will require building special instructions and instrumenting them into the application with the help of *use-def chains*.

Finally, an instrumented version of the assembly program is assembled and linked using the back-end of the same compiler, to generate the binary for the target architecture.

8.3.2   Architectural Design

Without losing the generality of our technique, we use an automatic processor design tool to implement the reliable processor model in hardware. This automatic design tool is used to design Application Specific Instruction-set Processors (ASIPs) [91, 92], custom designed for applications or application domains. Automatic processor design tools serves as a perfect starting place to build processor models.

Figure 8.5(b) describes the hardware design process of the model for the reliable processor. In the tool the functional units required to implement the processor will be chosen from a resource pool. Using the information of the ISAs, micro-instruction routines are formed, and are included into the processor model design. These micro-instruction

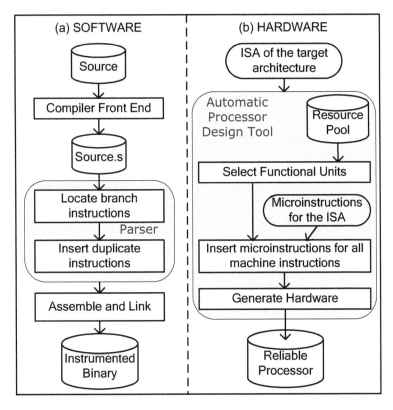

Figure 8.5: *Design flow for the control flow checking architecture*

routines will form the logic of the processor that will do the CFC along with regular operations at runtime. The final task in the architectural design process is to generate the hardware model in a hardware description language for simulation [behavioral] and synthesis [gate level] (indicated by *Reliable Processor* in Figure 8.5).

8.4   Implementation and Evaluation

Even though the techniques described in this chapter for CFC can be deployed in any type of embedded processor architecture, we have taken the PISA (portable instruction set architecture) instruction set as implemented in SimpleScalar™ tool set for our experimental implementation. The PISA instruction set is a simple MIPS like instruction set. CFC in the processor is enabled by altering the rapid processor design process described in [211] for hardware synthesis (allowing a processor described in VHDL which is synthesizable).

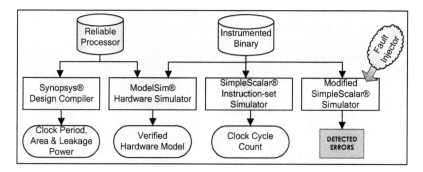

Figure 8.6: *Testing and Evaluation of the extended CFC*

To evaluate our approach, applications from MiBench benchmark suite were taken and compiled with the GNU/GCC® cross-compiler for the PISA instruction set. As mentioned previously, an automatic processor design tool, called ASIP Meister [293] is used to generate the VHDL description of the target processor as described in section 8.3.2. The output of the ASIP Meister are the VHDL models of the processor for simulation and synthesis.

As shown in Figure 8.6, the instrumented binary produced from the software design process and the processor simulation model generated by the hardware design tool are used in ModelSim® hardware simulator to verify the correctness of our design. Applications with different control flow instructions are simulated in Modelsim® and their behaviors are observed and verified by looking at the waveforms of related signals.

The gate level VHDL model from ASIP Meister is used with Synopsys Design Compiler® to obtain the clock period, area and leakage power overheads.

We evaluated the clock cycle overhead of the proposed architecture using a cycle accurate instruction set simulator, SimpleScalar™ 3.0/PISA tool set. The simulator is modified to include CFC as proposed in Section 8.2. The simulator is built around the existing cycle accurate simulator *sim-outorder* in the tool set and used to calculate the clock cycle overheads. As depicted in Figure 8.6, the same simulator is modified to perform fault injection analysis and the results are tabulated in the next Section. The micro-architectural parameters of SimpleScalar™ were configured to model a typical embedded processor as designed by ASIP meister. The parameters used for the simulated processor are shown in Table 8.1.

The clock cycle counts from the modified SimpleScalar™ simulator and the clock period from Synopsys Design Compiler® are used to calculate the total execution times of all the applications.

Parameter	Value	Parameter	Value
Issue	in-order	Issue width	1
Fetch queue size	4	Commit width	1
L1 I-cache	16kB	L1 D-cache	16kB
L1 I-cache latency	1 cycle	L1 D-cache latency	1 cycle
Initial memory latency	18 cycles	Memory latency	2 cycles

Table 8.1: *Architectural Parameters for Simulation of the extended CFC*

Parameters	Without CFC	With CFC	% overhead
Area (cells)	228489	236700	3.59
Clock Period (ns)	16.85	16.89	0.24
Leakage Power (μW)	485	503	3.71

Table 8.2: *Hardware Overhead of the extended CFC*

8.5   Experimental Results

In this section, we present memory, area and power overheads incurred by the proposed CFC solution, as well as the impact of the technique on performance (total execution time). Later we give results from the fault injection analysis performed on our model processor when our solution is implemented. For the purpose of experiments we have used the PISA instruction set as described in Section 8.4 and applications from the MiBench benchmark suite which represents typical workload for embedded processors.

8.5.1   Hardware Overhead

Table 8.2 tabulates the area, clock period and leakage power overheads due to the changes in the hardware. The overheads here represent the extra logic in our design. Taiwan Semiconductor Manufacturing Company's (TSMC) 90nm core library with typical conditions enabled is used for the hardware synthesis. The second column in Table 8.2 represents the parameters for the processor model without the CFC enabled (the base processor) and the third column represents the parameters of the processor model when the hardware for CFC is enabled. The percentage of overheads in area is 3.59%, clock period is 0.24% and leakage power is 3.71%.

8.5.2   Performance Overhead

Table 8.3 reports the performance overhead incurred by our scheme for different applications (first column) from MiBench benchmark suite. In Table 8.3 columns 2-4 tabulate the clock cycle comparisons (in millions) and percentage of overheads and columns 5-7 tabulate execution time comparison (in seconds) and percentage of overheads. The columns that are sub-titled *NoCFC* represent simulations of the processor model without the CFC enabled and *CFC* represents simulations of the processor model with the CFC enabled.

Benchmarks	Clock Cycle/10^6			Execution Time/s		
	NoCFC	CFC	%	NoCFC	CFC	%
adpcm.decode	121.6	122.8	0.99	2.05	2.07	1.23
adpcm.encode	89.96	90.12	0.18	1.52	1.52	0.42
blowf.encrypt	79.21	79.49	0.35	1.34	1.34	0.59
blowf.decrypt	80.44	81.05	0.76	1.37	1.37	1.00
crc32.checksum	57.62	57.62	0.00	0.97	0.97	0.24
jpeg.compress	16.41	16.56	0.91	0.28	0.28	1.15
jpeg.decompress	10.79	10.89	0.93	0.18	0.18	1.17
jpeg.transcoding	8.96	9.07	1.23	0.15	0.15	1.47

Table 8.3: *Performance Comparison with and without extended CFC*

From Table 8.3, the clock cycle overheads range from 0.00% to 1.23% with an average of 0.67% and the execution time overheads range from 0.24% to 1.47% with an average of 0.91%. In contrast, the performance (execution time) overhead in software only PECOS technique was in the range of 53.5-99.5%.

8.5.3   Codesize Overhead

Table 8.4 reports the code size overhead and the overhead in the number of instructions executed resulting from our software instrumentation for eight different applications (first column). In Table 8.4 columns 2-4 tabulate the comparison between the number of executed instructions (in millions) and percentage of overheads and columns 5-7 tabulate code size (the number of lines) comparisons and percentage of overheads. The columns with title *NoCFC* represents simulations of the processor model without the CFC enabled and *CFC* represents simulations of the processor model with the CFC enabled.

Benchmarks	Executed Inst./10^6			Code Size		
	NoCFC	CFC	%	NoCFC	CFC	%
adpcm.decode	76.55	92.98	13.88	623	676	8.50
adpcm.encode	58.91	72.60	15.37	618	670	8.40
blowf.encrypt	58.43	64.44	4.26	6463	6712	3.85
blowf.decrypt	59.04	65.05	5.97	6463	6712	3.85
crc32.checksum	42.51	50.72	12.82	527	549	4.23
jpeg.compress	11.62	12.88	7.41	58650	65160	11.1
jpeg.decompress	7.45	7.96	3.99	56577	62914	11.2
jpeg.transcoding	5.91	7.41	16.88	52605	58444	11.1

Table 8.4: *Codesize Comparisons with and without extended CFC*

From Table 8.4, overheads in the number of instructions executed ranges from 3.99% to 16.9% with an average of 10.7% and code size overhead ranges from 3.85% to 11.2% with an average of 7.78%. The code size overhead of software only PECOS technique was in the range of 50-150%.

8.5.4   Fault Injection Analysis

Table 8.5 tabulates the results of the fault injection analysis performed on our control flow detection architecture. Random memory errors are generated by performing bit bursts in the instruction memory. Random memory errors represent a wide range of transient errors in hardware and some errors in software. For each application, 10,000 faults are inserted one at a time and the runtime program behavior is observed to identify whether it is captured by our technique. With the current design, an average of 85.3% of the injected control flow errors are detected by our system. The rest of the erroneous control flow instructions fall into the library functions of the application which are presently not instrumented. We will be able to achieve 100% control flow error coverage, if we are able to instrument the library functions those are used in the applications.

Benchmarks	No. of Faults		CFEs	Captured by
	Total	Activated		our scheme
adpcm.decode	10000	5723	769	653
adpcm.encode	10000	4635	607	526
blowf.encrypt	10000	6242	911	824
blowf.decrypt	10000	6283	949	819
crc32.checksum	10000	6264	969	923
jpeg.compress	10000	2947	417	334
jpeg.decompress	10000	2901	428	342
jpeg.transcoding	10000	4176	710	558

Table 8.5: *Fault Injection Analysis for the extended CFC*

This fault injection analysis is only capable of testing our design for instruction memory bit flips and bursts detection. Further analysis could be performed to test the capability of the shadow register file and the shadow PC by injecting faults into the register file and the PC. However, all the injected errors in the register file will be captured by our scheme given that the control flow instructions which use the errornous registers are checked for errors. Therefore, the error coverage of the shadow register file is equal to the error coverage of the control flow instructions. Furthermore, all the injected errors in a PC will be captured by our scheme with 100% error coverage.

8.6   Future Work

The increasing number of multiple bit errors (bit bursts) in processors has increased the demand for solutions which covers multi bits error detection as opposed to single bit errors. The project described in this chapter and the previous chapter try to detect part of these errors in memory and special hardware components such as register-file. However, the need for solutions to cover other hardware components such as internal intermediate registers (pipeline registers for example) used in processors is still an open problem which need solutions. We identify this as a future work. Fault injection studies need to be performed to reveal the most vulnerable hardware components for transient faults in embedded processors. The results from such a study could enable selective replication and we will be able to improve the error coverage of our solution that we have proposed in this project. Therefore, we also identify such a fault injection study as a future work too.

8.7   Summary of this chapter

In this chapter, we have presented a hardware software technique to detect control flow errors caused by bit flips and bursts at runtime before an erroneous control flow instruction is executed. We have formulated a formal methodology to accommodate this technique within an automatic embedded processor design flow. Our evaluation studies reveal that the solution we have proposed is capable of handling control flow errors with as little as 3.59% of area and 0.91% of performance overheads. These overheads are minimal compared to the software solution that deals with the same problem. We conclude that by asserting CFC as a design requirement of an embedded processor, it is practicable to reduce the overhead of CFC as minimal as possible.

Fault injection analysis demonstrates that our solution is capable of capturing 85.3% of the injected control flow errors in the instruction memory. The limitation of the error coverage is due to the non instrumented code coming from the runtime libraries. Furthermore, our scheme will also detect control flow errors caused by bit flips and bursts in register file and PC with 100% coverage as long as the erroneous control flow instruction is instrumented. Our error detection scheme is preemptive and is capable of correcting control flow errors (assuming the error is in the original control flow instruction/register-file/PC) without any additional overheads. We believe that the technique described in this chapter could be used with any embedded processors for detecting control flow errors efficiently.

Chapter 9

Summary

... 'I wonder how many miles I've fallen by this time?' she said aloud. 'I must be getting somewhere near the centre of the earth. Let me see: that would be four thousand miles down, I think- yes that's about the right distance'...

— Lewis Carroll, *Alice's Adventures in Wonderland*

In this book, the authors have presented a framework, which was developed during the first author's research program in compliance with his PhD to solve security and reliability problems in embedded processors. The framework considers security and reliability as design parameters those need consideration during design time. Further, they have shown the advantages of this approach, as it overcomes the limitations of the current proposals, such as the performance and code size overheads in the software-only techniques and scalability problems in the hardware assisted techniques.

9.1 Future Work

Much work remains to be done. Individual chapters pointed out future research opportunities. This section highlights the most outstanding issues.

1. In the thesis presented in this book, the proposed security and reliability solutions for embedded processors have been implemented on a single issue processor. Recently, processor architectures to explore instruction level parallelism (ILP), such as superscalar[*] and VLIW[†](Very Long Instruction Word) are becoming common in

---

[*]Superscalar architectures attempt to speed up programs by reordering and/or executing instructions in parallel, using specialized and often complex hardware to discover these opportunities of ILP while the code executes [283].

[†]VLIW architectures execute instructions in parallel based on a fixed schedule determined when the code is compiled. They do not contain the specialized hardware associated with superscalar CPUs. Rather, they rely on compilers to analyze and schedule instructions in parallel. As a result, VLIW CPUs offer

embedded system market. Therefore, it is worth investigating the scalability of the proposed solutions in such architectures.

2. Further, scaling our framework on multi-processor based embedded systems will have its own challenges. This is also an area of interest for future work.

9.2 Final Remarks

Considering reliability and security issues during the design of an embedded system has its advantages as this approach overcomes the limitations of current proposals, such as the performance and code size overheads in the software-only techniques, and scalability problems in the hardware assisted techniques. Such a design time framework for embedded processors has been explored in this book. Further, the hardware and performance overheads, the error coverage and the latency of the error/intrusion detection of the systems proposed in this book have also been evaluated and shown to be minimal in comparison with other available methods.

---

significant computational power with less hardware complexity (but greater compiler complexity) than is associated with most superscalar CPUs [303].

Glossary

This glossary defines some of the terminology and acronyms that are used throughout the dissertation. Some of the concepts are defined in detail during the dissertation. Other concepts are assumed to be common knowledge. The terms are listed in alphabetical order.

AES   Advanced Encryption Standard

ALU   Arithmetic Logic Unit

ANSI   American National Standards Institute

API   Application Programming Interface

APWG   Anti Phishing Working Group

ASCII   American Standard Code for Information Interchange

ASIC   Application Specific Integrated Circuit

ASIP   Application Specific Instruction-set Processor

AST   Abstract Syntax Tree

BB   Basic Block

BTB   Branch Target Buffer

CC   Clock Cycle

CFC   Control Flow Checking

CFE	Control Flow Error
CFI	Control Flow Instruction
CFG	Control Flow Graph
CISC	Complex Instruction Set Computer
COTS	Commercial Off The Shelf
CPI	Clock Per Instruction
CPU	Central Processing Unit
CSI	Computer Security Institute
DES	Data Encryption Standard
DLL	Dynamic Link Libraries
DMAU	Data Memory Access Unit
DPA	Differential Power Analysis
DRAM	Dynamic Random Access Memory
ECB	Electronic Code Book
ECC	Error Correction Code
FBI	Federal Bureau on Investigation
FP	Frame Pointer
FPU	Floating Point Unit
FPGA	Field Programmable Gate Array
GCC	GNU C Compiler
GOT	Global Object Table
GPR	General Purpose Registers
GUI	Graphical User Interface
HDL	Hardware Description Language
HMM	Hidden Markov Model

ID	Instruction Decode
IDS	Intrusion Detection System
IF	Instruction Fetch
IR	Instruction Register
ILP	Instruction Level Parallelism
IMAU	Instruction Memory Access Unit
IPSec	Internet Protocol Security
ISA	Instruction Set Architecture
LAN	Local Area Network
MBU	Multi Bit Upset
MI	Micro-instruction
MISR	Multiple Input Shift Register
MPI	Message Passing Interface
MTBF	Mean Time Between Failures
PC	Program Counter
PCC	Portable C Compiler
PDA	Personal Digital Assistant
PLT	Procedure Linkage Table
RAM	Random Access Memory
RAS	Return Address Stack
RISC	Reduced Instruction Set Computer
RTL	Register Transfer Level
SBU	Single Bit Upset
SER	Soft Error Rate
SEU	Single Event Upset

SFI   Software based Fault Isolation

SP   Stack Pointer

SPA   Simple Power Analysis

SRAM   Static Random Access Memory

SRAS   Secure Return Address Stack

SSL   Secure Sockets Layer

SWIFI   SoftWare Implemented Fault Injection

VHDL   VHSIC Hardware Description Language

VHSIC   Very-High-Speed Integrated Circuit

VLIW   Very Long Instruction Word

WB   register Write Back

WEP   Wired Equivalent Privacy

WTLS   Wireless Transport Layer Security

Micro Instructions

This appendix describes the micro instructions used throughout this dissertation.

B.1 Micro-instructions Grammar

This section provides the grammar for the micro-instruction used in this dissertation in *BackusNaur* Form.

<Program> ::= {<VarDeclaration> <MicroInstructionDesc>}

<VarDeclaration> ::= { <WireDeclaration> }

<WireDeclaration> ::= **wire** <BitRangeSpec> <WireName> ;

<WireName> ::= <Id>

<Id> ::= <Alphabet> {<AlphaNumeral>}

<Alphabet> ::= **[a-zA-Z]**

<AlphaNumeral> ::= <Alphabet> | <Number>

<Number> ::= **[0-9]**

<MicroInstructionDesc> ::= <VarDeclaration> <Statements>

<Statements> ::= {<Statement>}

<Statement> ::= <SimpleAssignStatement> | <ConditionalAssignStatement>
| <ConditionalExecutionStatement>

<SimpleAssignStatement> ::= <LeftSide> **=** <RightSide> **;**

<LeftSide> ::= <VariableName> | <VariableNameSet> | **null**

<RightSide> ::= <BitWiseAND> | <BitWiseOR> | <OnesComplement>
| <Comparison> | <Sets> | <ResourceReference> |
<PartialSelect> | <Constant> | <VariableReference>

<BitWiseAND> ::= <VariableReference> **AND** <VariableReference>

<BitWiseOR> ::= <VariableReference> **OR** <VariableReference>

<OnesComplement> ::= **NOT** <VariableReference>

<Comparison> ::= <VariableReference> <RelationalOperator> <Constant>

<RelationalOperator> ::= **==** | **!=**

<VariableReference> ::= <VariableName>

<VariableName> ::= <WireName>

<VariableNameSet> ::= **<** <VariableName> {**,** <VariableName>} **>**

<Sets> ::= **<** <VariableReference> {**,** <VariableReference>} **>**

- <ResourceReference> ::= <ResourceName> **.** <FunctionName>
(**[** <ArgumentList> **)**

- <FunctionName> ::= <Id>

- <ArgumentList> ::= <Argument> { **,** <Argument> }

- <Argument> ::= <VariableReference>

<ConditionalAssignStatement> ::= <LeftSide> **=** **(** <VariableReference> **)** **?** <VariableReference>
: <VariableReference> **;**

<ConditionalExecutionStatement> ::= <LeftSide> **=** **[** <VariableReference> **]** <ResourceFunSpecifi
;

<ResourceFunSpecification> ::= <ResourceName>**.**<FunctionName> **(** <ArgumentList>
)

<Constant> ::= <BitLiteral> | <VectorLiteral>

<BitLiteral> ::= **"** <BinaryCharacter> {<BinaryCharacter>} **"**

<BinaryCharacter> ::= **0** | **1**

B.2   Elements of Micro-instructions

This section details different type of elements used in the micro-instructions.

B.2.1   Variable

In micro-instructions variables are used when transferring the data. In order to use the variable, it must be declared before its use. Refer to the grammar to see the detailed information for variable declaration.

B.2.2   Constant

All the constants used for micro-operation description shall be binary constants. The binary constants may be grouped in two. First group is 'bit literal', which is represented by 0 (zero bracketed with single quote) or 1(one bracketed with single quote). Second is 'vector literal'. It is represented by sequence of 0 and 1 bracketed with double quote (ex. 010). Generally, one bit constant is represented by bit literal, and constants longer than two bits are represented by vector literal.

B.2.3   Bitwise Operator

The grammar represents three bitwise operators and they are AND, OR and NOT. All values of bitwise operators should be variables.

B.2.4   Concatenation

When you concatenate bits, you are required to use brackets '<>' to show the concatenation. Variables are the only available element that you can concatenate. For example, if you want to concatenate three variables a(1 bit), b(3 bit), and c(4 bit) to handle as one variable d(8 bit), write the following statement.

d=<a,b,c>;

Here the most significant bit of variable d is that of the variable a, and the least significant bit of variable d is that of variable c.

B.2.5 Relational Operator

Relational operators are used to compare the variable and provided constants. There are two relational operators and they are equality and inequality. For example:

b=a==01;

b becomes 1 when a is equal to 01. If a is not equal to 01, b becomes as 0.

B.2.6 Conditional Operation

Conditional operations could be grouped into two categories. The first one is conditional data transfer. This is written as follows:

result = (condition) ? a:b;

The same concept of ternary operator of C programming language applies to this statement. For this example when conditional variable condition is 1, assign a to result, when 0 assign b to result.

The second is conditional resource execution. The following example shows how the write function of resource REG is executed, when conditional variable *condition* is set to 1.

null = [condition] REG.write(data);

Bibliography

[1] Aleph One. Smashing the Stack for Fun and Profit. *Phrack Magazine*, 7(49), 1996. (Cited on pages 17.)

[2] Z. Alkhalifa, V. Nair, N. Krishnamurthy, and J. Abraham. Design and evaluation of system-level checks for on-line control flow error detection. *IEEE Transactions on Parallel and Distributed Systems*, 10(6):627–641, June 1999. (Cited on pages 78, 79.)

[3] S. Almukhaizim, P. Drineas, and Y. Makris. On compaction-based concurrent error detection. In *Proceedings of the 9th ieee international on-line testing symposium (iolts 2003)*, page 157. IEEE Computer Society, 2003. (Cited on pages 74.)

[4] A. Alomary, T. Nakata, Y. Honma, and J. Sato. PEAS-I: a hardware/software co-design system for asips. *Proceedings of the IEEE International Test Conference*, pages 2–7, 1993. (Cited on pages 88.)

[5] Anonymous. Once upon a free(). *Phrack Magazine*, 11(57), 2001. (Cited on pages 19.)

[6] Anti-Phishing Working Group. Phishing Activity Trends Report. Technical report, Available at http://www.antiphishing.org/resources.html, 2005. (Cited on pages 13, 15.)

[7] APWG. Anti-Phishing Working Group. Available at http://www.antiphishing.org/. (Cited on pages 13.)

[8] J. Arlat, M. Aguera, L. Amat, Y. Crouzet, J.-C. Fabre, J.-C. Laprie, E. Martins, and D. Powell. Fault injection for dependability validation: A methodology and some applications. *IEEE Transactions on Software Engineering*, 16(2):166–182, 1990. (Cited on pages 83.)

[9] J. Arlat, M. Aguera, Y. Crouzet, J. Fabre, E. Martins, and D. Powell. Experimental Evaluation of the Fault Tolerance of an Atomic Multicast System. *IEEE Transactions on Reliability*, 39(4):455–467, 1990. (Cited on pages 83.)

169

[10] J. Arlat, Y. Crouzet, and J.-C. Laprie. Fault injection for dependability validation of fault-tolerant computing systems. In *Proceedings of the 19th International
Symposium on Fault-Tolerant Computing*, pages 348–355, 1989. (Cited on pages 83.)

[11] D. Arora, S. Ravi, A. Raghunathan, and N. K. Jha. Secure embedded processing
through hardware-assisted runtime monitoring. In *Proceedings of the Design, Automation and Test in Europe (DATE'05)*, volume 1, 2005. (Cited on pages 3, 43, 44,
111.)

[12] D. Arora, S. Ravi, A. Raghunathan, M. Sankaradass, N. Jha, and S. T. Chakradhar.
Software Architecture Exploration for High-Performance Security Processing on a
Multiprocessor Mobile SoC. In *Proceedings of the Design and Automation Conference 2006 (DAC'06)*, pages 496–501, San Fransisco, California, USA, November
2006. ACM Press. (Cited on pages 2.)

[13] K. Ashcraft and D. Engler. Using programmer-written compiler extensions to catch
security holes. In *Proceedings of the 2002 ieee symposium on security and privacy
(Sp '02)*, page 143, Washington, DC, USA, 2002. IEEE Computer Society. (Cited
on pages 46.)

[14] T. M. Austin, S. E. Breach, and G. S. Sohi. Efficient detection of all pointer and
array access errors. In *Proceedings of the acm sigplan 1994 conference on programming language design and implementation (Pldi '94)*, pages 290–301, New
York, NY, USA, 1994. ACM Press. (Cited on pages 63, 64, 65.)

[15] A. Avizienis and J. Kelly. Fault Tolerance by Design Diversity: Concepts and
Experiments. *IEEE Computer (magazine)*, 17:67–80, 1984. (Cited on pages 73.)

[16] A. Avizienis, J. Laprie, and B. Randell. Fundamental concepts of dependability,
2001. (Cited on pages 11.)

[17] A. Avizienis, J.-C. Laprie, B. Randell, and C. Landwehr. Basic concepts and tax-
onomy of dependable and secure computing. *IEEE Transactions on Dependable
and Secure Computing*, 1:11–33, Jan.-March 2004. (Cited on pages 11.)

[18] D. R. Avresky, J. Arlat, J.-C. Laprie, and Y. Crouzet. Fault injection for the formal
testing of fault tolerance. *IEEE Transactions on Reliability*, 45(3):443–455, 1996.
(Cited on pages 83.)

[19] S. Bagchi, Y. Liu, K. Whisnant, Z. Kalbarczyk, R. K. Iyer, Y. Levendel, and
L. Votta. A framework for database audit and control flow checking for a wireless
telephone network controller. In *Proceedings of the 2001 international conference*

on dependable systems and networks (DSN'01), pages 225–234, Washington, DC, USA, 2001. IEEE Computer Society. (Cited on pages 79.)

[20] A. Baratloo, N. Singh, and T. Tsai. Transparent run-time defense against stack smashing attacks. In *Proceedings of 9th USENIX Security Symposium*, June 2000. (Cited on pages 67, 68.)

[21] E. G. Barrantes, D. H. Ackley, T. S. Palmer, D. Stefanovic, and D. D. Zovi. Randomized instruction set emulation to disrupt binary code injection attacks. In *Proceedings of the 10th acm conference on computer and communications security (Ccs '03)*, pages 281–289, New York, NY, USA, 2003. ACM Press. (Cited on pages 43.)

[22] J. F. Bartlett. A nonstop kernel. In *Proceedings of the eighth acm symposium on operating systems principles (Sosp'81)*, pages 22–29, New York, NY, USA, 1981. ACM Press. (Cited on pages 81.)

[23] J. H. Barton, E. W. Czeck, Z. Z. Segall, and D. P. Siewiorek. Fault injection experiments using fiat. *IEEE Transactions on Computers*, 39(4):575–582, 1990. (Cited on pages 84.)

[24] R. Baumann. Radiation-induced soft errors in advanced semiconductor technologies. *IEEE Transactions on Device and Materials Reliability*, 5(3):305–316, 2005. (Cited on pages 6, 7, 8.)

[25] G. Begic. An Introduction to Runtime Analysis with Rational PurifyPlus. Technical report, IBM, 2003. (Cited on pages 50.)

[26] A. Benso, M. Rebaudengo, and M. S. Reorda. Flexfi: a flexible fault injection environment for microprocessor-based systems. In *Proceedings of the 18th International Conference on Computer Computer Safety, Reliability and Security*, pages 323–335, 1999. (Cited on pages 85.)

[27] S. Bhatkar, D. C. DuVarney, and R. Sekar. Address obfuscation: an efficient approach to combat a broad range of memory error exploits. In *Proceedings of the 12th USENIX security symposium*, Washington, DC, August 2003. (Cited on pages 70.)

[28] blexim. Basic integer overflows. *Phrack Magazine*, 11(60), December 2002. (Cited on pages 25.)

[29] D. Boneh, R. A. DeMillo, and R. J. Lipton. On the importance of checking cryptographic protocols for faults. *Lecture Notes in Computer Science*, 1233:37–51, 1997. (Cited on pages 12, 100.)

[30] D. Boneh, R. A. DeMillo, and R. J. Lipton. On the importance of eliminating errors in cryptographic computations. *Journal of Cryptology*, 14(2):101–119, 2001. (Cited on pages 12.)

[31] E. Borin, C. Wang, Y. Wu, and G. Araujo. Dynamic binary control-flow errors detection. *SIGARCH Computer Architecture News*, 33(5):15–20, 2005. (Cited on pages 80, 81.)

[32] N. S. Bowen and D. K. Pradhan. Processor- and memory-based checkpoint and rollback recovery. *IEEE Computer (magazine)*, 26(2):22–31, 1993. (Cited on pages 82.)

[33] B. Bray. Compiler Security Checks In Depth. *Available at http://www.codeproject.com/tips/seccheck.asp*, February 2002. (Cited on pages 61.)

[34] Bruce Schneier. *Applied Cryptography: Protocols, Algorithms, and Source Code in C*. John Wiley and Sons, second edition edition, 1996. (Cited on pages 3, 113.)

[35] D. Brumley and D. Boneh. Remote timing attacks are practical. In *Proceedings of the 12th USENIX Security Symposium*, August 2003. (Cited on pages 12.)

[36] Bulba and Kil3r. Bypassing Stackguard and Stackshield. *Phrack Magazine*, 10(56), 2000. (Cited on pages 18, 60, 62.)

[37] D. Burger and T. M. Austin. The simplescalar tool set, version 2.0. *SIGARCH Computer Architecture News*, 25(3):13–25, 1997. (Cited on pages 103, 123, 139.)

[38] W. R. Bush, J. D. Pincus, and D. J. Sielaff. A static analyzer for finding dynamic programming errors. *Software: Practice and Experience*, 30(7):775–802, 2000. (Cited on pages 46.)

[39] J. Carreira, H. Madeira, and J. Silva. Xception: Software fault injection and monitoring in processor functional units. In *Proceedings of the 5th International Working Conference on Dependable Computing for Critical Applications (DCCA-5)*, pages 135–149, Urbana-Champaign, IL, USA, 1995. (Cited on pages 84.)

[40] J. Carreira, H. Madeira, and J. G. Silva. Xception: A technique for the experimental evaluation of dependability in modern computers. *IEEE Transactions on Software Engineering*, 24(2):125–136, 1998. (Cited on pages 84.)

[41] C. S. Cates. Where's waldo: Uncovering hard-to-find application killers. In *Proceedings of the 30th International Computer Measurement Group Conference, December 5-10, 2004, Las Vegas, Nevada, USA*, pages 727–740. Computer Measurement Group, 2004. (Cited on pages 50.)

[42] CERT Coordination Center. Vulnerability Notes Database, CERT Coordination Center, 2004. (Cited on pages 15.)

[43] CERT Coordination Center. CERT/CC Vulnerabilities Statistics 1988-2005, CERT Coordination Center, 2005. (Cited on pages 15.)

[44] CERT Coordination Center. CERT/CC Vulnerabilities Statistics 1988-2006, CERT Coordination Center, 2006. (Cited on pages 2, 3, 4.)

[45] S. Cesare. ELF Executable Reconstruction from a Core Image. Available at http://vx.netlux.org/lib/vsc03.html, December 1999. (Cited on pages 29.)

[46] R. Chandra, M. Cukier, R. M. Lefever, and W. H. Sanders. Loki: A state-driven fault injector for distributed systems. In *Proceedings of the International Conference on Dependable Systems and Networks (DSN-2000), New York, NY*, pages 237–242, 2000. (Cited on pages 84.)

[47] Charles Price. MIPS IV Instruction Set. MIPS Technologies, Inc. (revision 3.1), January 1995. (Cited on pages 103.)

[48] M. Chew and D. Song. Mitigating buffer overflows by operating system randomization. Technical Report CMU-CS-02-197, Department of Computer Science, Carnegie Mellon University, December 2002. (Cited on pages 71.)

[49] T.-c. Chiueh and F.-H. Hsu. RAD: a compile-time solution to buffer overflow attacks. In *Proceedings of the the 21st international conference on distributed computing systems (Icdcs '01)*, page 409, Washington, DC, USA, 2001. IEEE Computer Society. (Cited on pages 60, 61.)

[50] G. S. Choi and R. K. Iyer. Focus: An experimental environment for fault sensitivity analysis. *IEEE Transactions on Computers*, 41(12):1515–1526, 1992. (Cited on pages 84.)

[51] J. Clark and D. Pradhan. REACT: a synthesis and evaluation tool for fault-tolerant multiprocessor architectures. In *Proceedings in the Annual Reliability and Maintainability Symposium, 1993.*, pages 428 – 435, 1993. (Cited on pages 84.)

[52] J. A. Clark and D. K. Pradhan. Fault injection. *IEEE Computer (magazine)*, 28(6):47–56, 1995. (Cited on pages 84.)

[53] J. Condit, M. Harren, S. McPeak, G. C. Necula, and W. Weimer. CCured in the real world. In *Proceedings of the acm sigplan 2003 conference on programming language design and implementation*, pages 232–244, New York, NY, USA, 2003. ACM Press. (Cited on pages 45.)

[54] J. Connet, E. Pasternak, and B. Wagner. Software Defenses in Real Time Control Systems. In *Proceedings of the 2nd International Symposium on Fault Tolerant Computing*, pages 94–99, 1972. (Cited on pages 72.)

[55] P. Cousot and N. Halbwachs. Automatic discovery of linear restraints among variables of a program. In *Proceedings of the 5th acm sigact-sigplan symposium on principles of programming languages (Popl '78)*, pages 84–96, New York, NY, USA, 1978. ACM Press. (Cited on pages 46.)

[56] C. Cowan, M. Barringer, S. Beattie, G. Kroah-Hartman, M. Frantzen, and J. Lokier. Formatguard: Automatic protection from printf format string vulnerabilities. In *Proceedings of the 10th USENIX Security Symposium*, pages 191–200. USENIX Association, 2001. (Cited on pages 67.)

[57] C. Cowan, S. Beattie, R. F. Day, C. Pu, P. Wagle, and E. Walthinsen. Protecting Systems from Stack Smashing Attacks with StackGuard. In *Proceedings of Linux Expo 1999*, 1999. (Cited on pages 61.)

[58] C. Cowan, S. Beattie, J. Johansen, and P. Wagle. PointGuard: protecting pointers from buffer overflow vulnerabilities. In *Proceedings of the 12th USENIX Security Symposium*, pages 91–104, Washington, District of Columbia, U.S.A, August 2003. USENIX Association. (Cited on pages 62, 63, 64.)

[59] C. Cowan, C. Pu, D. Maier, J. Walpole, P. Bakke, S. Beattie, A. Grier, P. Wagle, Q. Zhang, and H. Hinton. StackGuard: Automatic Adaptive Detection and Prevention of Buffer-Overflow Attacks. In *Proceedings of the 7th USENIX Security Conference*, pages 63–78, San Antonio, Texas, 1998. (Cited on pages 61, 62.)

[60] M. Cukier, R. Chandra, D. Henke, J. Pistole, and W. H. Sanders. Fault Injection based on a Partial view of the Global State of a Distributed System. In *Proceedings of the 18th IEEE Symposium on Reliable Distributed Systems (SRDS '99)*, page 168, Washington, DC, USA, 1999. IEEE Computer Society. (Cited on pages 84.)

[61] E. W. Czeck. *On the prediction of fault behavior based on workload*. PhD thesis, Carnegie Mellon University, Pittsburgh, PA, USA, 1991. (Cited on pages 84.)

[62] C. da Lu and D. A. Reed. Assessing fault sensitivity in mpi applications. In *Proceedings of the 2004 acm/ieee conference on supercomputing*, page 37, Washington, DC, USA, 2004. IEEE Computer Society. (Cited on pages 32.)

[63] S. F. Daniels. A concurrent test technique for standard microprocessors. In *Digest of Papers COMPCON Spring 83*, pages 389–394, San Francisco, CA, 1983. (Cited on pages 72.)

[64] S. Dawson, F. Jahanian, T. Mitton, and T.-L. Tung. Testing of fault-tolerant and real-time distributed systems via protocol fault injection. In *Proceedings of the the twenty-sixth annual international symposium on fault-tolerant computing (ftcs '96)*, page 404, Washington, DC, USA, 1996. IEEE Computer Society. (Cited on pages 84.)

[65] P. Deeprasertkul, P. Bhattarakosol, and F. O'Brien. Automatic detection and correction of programming faults for software applications. *Journal of Systems and Software*, 78(2):101–110, 2005. (Cited on pages 52.)

[66] R. DeLine and M. Fahndrich. Enforcing high-level protocols in low-level software. *ACM SIGPLAN Notices*, 36(5):59–69, 2001. (Cited on pages 45.)

[67] X. Delord and G. Saucier. Control flow checking in pipelined RISC microprocessors: the Motorola MC88100 case study. In *Proceedings of the euromicro '90 workshop on real time*, pages 162–169. IEEE Computer Society, June 1990. (Cited on pages 75.)

[68] X. Delord and G. Saucier. Formalizing Signature Analysis for Control Flow Checking of Pipelined RISC Microprocessors. In *Proceedings of international test conference*, pages 936–945, October 1991. (Cited on pages 75.)

[69] DES Definition. Wikipedia Foundation Inc., the free encyclopedia,. http://en.wikipedia.org/wiki/Data_Encryption_Standard, 2006. (Cited on pages 121.)

[70] S. Designer. Non-executable stack patch. Avaliable at http://www.usenix.org/events/sec02/full_papers/lhee/lhee_html/node7.html, 1997. (Cited on pages 71.)

[71] S. Designer. JPEG COM Marker Processing Vulnerability in Netscape Browsers. *Bugtraq*, July 2000. (Cited on pages 19.)

[72] D. Dhurjati and V. Adve. Backwards-compatible array bounds checking for c with very low overhead. In *Proceedings of the 28th international conference on software engineering (Icse '06)*, pages 162–171, New York, NY, USA, 2006. ACM Press. (Cited on pages 64.)

[73] D. Dhurjati, S. Kowshik, V. Adve, and C. Lattner. Memory safety without runtime checks or garbage collection. In *Proceedings of the 2003 acm sigplan conference on language, compiler, and tool for embedded systems (Lctes '03)*, pages 69–80, New York, NY, USA, 2003. ACM Press. (Cited on pages 45.)

[74] I. Dobrovitski. Exploit for CVS double free() for Linux pserver. Bugtraq: Available at http://seclists.org/lists/bugtraq/2003/Feb/0042.html, February 2003. (Cited on pages 22.)

[75] N. Dor, M. Rodeh, and M. Sagiv. CSSV: towards a realistic tool for statically detecting all buffer overflows in c. In *Proceedings of the acm sigplan 2003 conference on programming language design and implementation (Pldi '03)*, pages 155–167, New York, NY, USA, 2003. ACM Press. (Cited on pages 46.)

[76] N. Dor, M. Rodeh, and S. Sagiv. Cleanness checking of string manipulations in c programs via integer analysis. In *Proceedings of the 8th international symposium on static analysis*, pages 194–212, London, UK, 2001. Springer-Verlag. (Cited on pages 46.)

[77] DRAM Definition. Wikipedia Foundation Inc., the free encyclopedia,. http://en.wikipedia.org/wiki/Dynamic_random_access_memory, 2006. (Cited on pages 7.)

[78] J. G. Dyer, M. Lindemann, R. Perez, R. Sailer, L. van Doorn, S. W. Smith, and S. Weingart. Building the ibm 4758 secure coprocessor. *IEEE Computer (magazine)*, 34(10):57–66, 2001. (Cited on pages 36, 37.)

[79] J. Eifert and J. Shen. Processor Monitoring Using Asynchronous Signatured Instruction Streams. In *Proceedings of the International Symposium of Fault-Tolerant Computing*, pages 394–399, 1984. (Cited on pages 75.)

[80] Ü. Erlingsson and F. B. Schneider. SASI enforcement of security policies: a retrospective. In *Proceedings of the 1999 workshop on New security paradigms (NSPW '99)*, pages 87–95, New York, NY, USA, 2000. ACM Press. (Cited on pages 56, 57.)

[81] B. Eschermann. On combining off-line BIST and on-line control flow checking. In *Proceedings of the Twenty-second international symposium on fault-tolerant computing, 1992. ftcs-22*, pages 298–305. IEEE Computer Society, July 1992. (Cited on pages 74.)

[82] H. Etoh and K. Yoda. Protecting from stack-smashing attacks. Report, IBM Research Division, Tokyo Research Laboratory, 1623-14 Shimotsuruma, Yamato, Kanagawa 242-8502, Japan, 2000. (Cited on pages 62.)

[83] D. Evans and D. Larochelle. Improving security using extensible lightweight static analysis. *IEEE Transactions on Software Engineering*, 19(1):42–51, 2002. (Cited on pages 45.)

[84] D. Evans and A. Twyman. Flexible policy-directed code safety. In *IEEE Symposium on Security and Privacy*, pages 32–45, 1999. (Cited on pages 57, 58.)

[85] H. Eveking. Superscalar DLX Processor, 2001. (Cited on pages 107.)

[86] C. Fetzer and Z. Xiao. Detecting heap smashing attacks through fault containment wrappers. In *Proceedings of the 20th Symposium on Reliable Distributed Systems (SRDS 2001)*, pages 80–89, 2001. (Cited on pages 68.)

[87] C. Fetzer and Z. Xiao. An automated approach to increasing the robustness of c libraries. In *Proceedings of the 2002 International Conference on Dependable Systems and Networks (DSN 2002)*, pages 155–166, 2002. (Cited on pages 68, 69.)

[88] C. Fetzer and Z. Xiao. A flexible generator architecture for improving software dependability. In *Proceedings of the 13th International Symposium on Software Reliability Engineering (ISSRE 2002)*, pages 102–116, 2002. (Cited on pages 68, 69.)

[89] C. Fetzer and Z. Xiao. Healers: A toolkit for enhancing the robustness and security of existing applications. In *Proceedings of the 2003 International Conference on Dependable Systems and Networks (DSN 2003)*, pages 317–322, 2003. (Cited on pages 68, 70.)

[90] G. Fink and M. Bishop. Property-based testing: a new approach to testing for assurance. *SIGSOFT Software Engineering Notes*, 22(4):74–80, 1997. (Cited on pages 51.)

[91] J. A. Fisher. Customized instruction-sets for embedded processors. In *Proceedings of the 36th ACM/IEEE conference on Design Automation (DAC'99)*, pages 253–257. ACM Press, 1999. (Cited on pages 102, 137, 151.)

[92] J. A. Fisher, P. Faraboschi, and G. Desoli. Custom-fit processors: letting applications define architectures. In *Proceedings of the 29th annual acm/ieee international symposium on microarchitecture (Micro 29)*, pages 324–335. IEEE Computer Society, 1996. (Cited on pages 102, 151.)

[93] S. Forrest, S. A. Hofmeyr, A. Somayaji, and T. A. Longstaff. A sense of self for unix processes. In *Proceedings of the 1996 ieee symposium on security and privacy*, page 120, Washington, DC, USA, 1996. IEEE Computer Society. (Cited on pages 53.)

[94] J. S. Foster, M. Fähndrich, and A. Aiken. A theory of type qualifiers. In *Proceedings of the 1999 ACM SIGPLAN Conference on Programming Language Design and Implementation (PLDI)*, pages 192–203, 1999. (Cited on pages 46.)

[95] Franklin M.J. A study of time redundant fault tolerant techniques for superscalar processors. *Proceedings of IEEE International Workshop on Defect and Fault Tolerannce in VLSI Systems*, 1995. (Cited on pages 82.)

[96] M. Frantzen and M. Shuey. StackGhost: hardware facilitated stack protection. In *Proceedings of the 10th USENIX Security Symposium*, pages 55–66, 2001. (Cited on pages 72.)

[97] J. Gaisler. Concurrent error-detection and modular fault-tolerance in a 32-bit processing core for embedded space flight applications. In *Proceedings of the the Twenty-Fourth Annual International Symposium on Fault-Tolerant Computing (FTCS/24)*, pages 128–130, 1994. (Cited on pages 75, 76, 134, 146.)

[98] V. Ganapathy, S. Jha, D. Chandler, D. Melski, and D. Vitek. Buffer overrun detection using linear programming and static analysis. In *Proceedings of the 10th ACM conference on computer and communications security*, pages 345–354, New York, NY, USA, 2003. ACM Press. (Cited on pages 49.)

[99] C. H. Gebotys. Low energy security optimization in embedded cryptographic systems. In *Proceedings of the 2nd ieee/acm/ifip international conference on hardware/software codesign and system synthesis (Codes+isss '04)*, pages 224–229. ACM Press, 2004. (Cited on pages 37.)

[100] A. Ghosh, B. Johnson, and J. Profeta. System-level modeling in the adept environment of a distributed computer system for real-time applications. In *Proceedings of the international computer performance and dependability symposium on computer performance and dependability symposium (Ipds '95)*, page 194, Washington, DC, USA, 1995. IEEE Computer Society. (Cited on pages 84.)

[101] A. K. Ghosh and T. O'Connor. Analyzing programs for vulnerability to buffer overrun attacks. In *Proceedings of the 21st NIST-NCSC national information systems security conference*, pages 274–382, 1998. (Cited on pages 51.)

[102] A. K. Ghosh, T. O'Connor, and G. McGraw. An automated approach for identifying potential vulnerabilities in software. In *Proceedings of the 1998 IEEE Symposium on Security and Privacy*, pages 104–114, Oakland, CA, 1998. (Cited on pages 51.)

[103] I. Goldberg, D. Wagner, R. Thomas, and E. A. Brewer. A Secure Environment for Untrusted Helper Applications. In *Proceedings of the 6th USENIX Security Symposium*, San Jose, CA, USA, 1996. (Cited on pages 58.)

[104] O. Goloubeva, M. Rebaudengo, M. Reorda, and M. Violante. Improved software-based processor control-flow errors detection technique. In *Proceedings of the Annual Reliability and Maintainability Symposium*, pages 583–589. IEEE Computer Society, January 2005. (Cited on pages 78.)

[105] O. Goloubeva, M. S. Rebaudengo, M. S. Reorda, and M. Violante. Soft-error detection using control flow assertions. In *Proceedings of the 18th ieee international symposium on defect and fault-tolerance in vlsi systems (DFT 2003)*, pages 581–588, November 2003. (Cited on pages 78.)

[106] L. A. Gordon, M. P. Loeb, W. Lucyshyn, and R. Richardson. CSI/FBI Computer Crime and Security Survey. Technical Report 11, Computer Security Institute, San Francisco, CA, USA, 2006. (Cited on pages 2, 4, 5.)

[107] K. Goswami and R. Iyer. DEPEND: A Design Environment for Prediction and Evaluation of System Dependability. In *Proceedings of the Ninth Digital Avionics Systems Conference*, 1990. (Cited on pages 84.)

[108] K. K. Goswami, R. K. Iyer, and L. Young. Depend: A simulation-based environment for system level dependability analysis. *IEEE Transactions on Computers*, 46(1):60–74, 1997. (Cited on pages 84.)

[109] R. Grimes. Preventing Buffer Overflows in C++. *Dr Dobb's Journal: Software Tools for the Professional Programmer*, 29(1):49–52, 2004. (Cited on pages 61.)

[110] D. Grossman, M. Hicks, T. Jim, and G. Morrisett. Cyclone: a type-safe dialect of C. *C/C++ Users Journal*, 23(1), January 2005. (Cited on pages 44.)

[111] D. Grossman, J. G. Morrisett, T. Jim, M. W. Hicks, Y. Wang, and J. Cheney. Region-based memory management in cyclone. In *Proceedings of the 2002 ACM SIGPLAN Conference on Programming Language Design and Implementation (PLDI)*, pages 282–293. ACM Press, 2002. (Cited on pages 44.)

[112] C. S. Guenzer, E. Wolicki, and R. Allas. Single Even Upsets of dynamic RAMs by neutron and protons. *IEEE Transactions on Nuclear Science*, 26(5048):1485–1489, 1979. (Cited on pages 6.)

[113] U. Gunneflo, J. Karlsson, and J. Torin. Evaluation of Error Detection Schemes Using Fault Injection by Heavy-Ion Radiation. In *Proceedings of IEEE International Symposium on Fault-Tolerant Computing*, pages 340–347, 1989. (Cited on pages 83.)

[114] Gunter Ollmann. Second-order Code Injection Attacks: Advanced Code Injection Techniques and Testing Procedures, 2003. (Cited on pages 111.)

[115] M. R. Guthaus, J. S. Ringenberg, D. Ernst, T. M. Austin, T. Mudge, and R. B. Brown. Mibench: A free, commercially representative Embedded Benchmark Suite. *Proceedings of IEEE 4th Annual Workshop on Workload Characterization, Austin, TX*, pages 83–94, December 2001. (Cited on pages 92, 110, 132.)

[116] J. Guthoff. Combining software-implemented and simulation-based fault injection into a single fault injection method. In *Proceedings of the twenty-fifth international symposium on fault-tolerant computing*, page 196, Washington, DC, USA, 1995. IEEE Computer Society. (Cited on pages 85.)

[117] R. Hastings and B. Joyce. Purify: A Tool for Detecting Memory Leaks and Access Errors in C and C++ Programs. In *Proceedings of the USENIX Winter 1992 Conference*, pages 125 – 138, San Francisco, CA, 1992. Pure Software Inc., USENIX. (Cited on pages 47, 50.)

[118] E. Haugh and M. Bishop. Testing c programs for buffer overflow vulnerabilities. In *Proceedings of the Network and Distributed System Security Symposium (NDSS 2003)*, 2003. (Cited on pages 50.)

[119] E. Hess, N. Janssen, B. Mayer, and T. Schutze. Information leakage attacks agaist smart card implementations of cryptographic algorithms and countermeasures. In *Proceedings of eurosmart security conference*, pages 55–64, June 2000. (Cited on pages 100.)

[120] M. Hicks, G. Morrisett, D. Grossman, and T. Jim. Experience with safe manual memory-management in cyclone. In *Proceedings of the 4th international symposium on memory management (Ismm '04)*, pages 73–84, New York, NY, USA, 2004. ACM Press. (Cited on pages 44.)

[121] S. A. Hofmeyr, S. Forrest, and A. Somayaji. Intrusion detection using sequences of system calls. *Journal of Computer Security*, 6(3):151–180, 1998. (Cited on pages 52.)

[122] M.-C. Hsueh, T. K. Tsai, and R. K. Iyer. Fault injection techniques and tools. *IEEE Computer (magazine)*, 30(4):75–82, 1997. (Cited on pages 32.)

[123] J. Irwin, D. Page, and N. P. Smart. Instruction stream mutation for non-deterministic processors. In *Proceedings of the ieee international conference on application-specific systems, architectures, and processors (Asap '02)*, page 286, Washington, DC, USA, 2002. IEEE Computer Society. (Cited on pages 12.)

[124] V. S. Iyengar and L. L. Kinney. Concurrent fault detection in microprogrammed control units. *IEEE Transactions on Computers*, 34(9):810–821, 1985. (Cited on pages 72.)

[125] R. K. Iyer, N. M. Nakka, Z. T. Kalbarczyk, and S. Mitra. Recent advances and new avenues in hardware-level reliability support. *IEEE Micro*, 25(6):18–29, 2005. (Cited on pages 6, 77.)

[126] E. Jenn, J. Arlat, M. Rimén, J. Ohlsson, and J. Karlsson. Fault injection into VHDL models: the MEFISTO tool. In *Digest of Papers of the Twenty-Fourth International Symposium on Fault-Tolerant Computing (FTCS-24)*, volume Vol., Iss., 15-17, pages 66–75, June 1994. (Cited on pages 84.)

[127] N. Jha. Fault-tolerant computer system design [Book Reviews]. *IEEE Transaction on Parallel and Distributed Technology: Systems and Applications*, 4(4):84, 1996. (Cited on pages 82.)

[128] T. Jim, G. Morrisett, D. Grossman, M. Hicks, J. Cheney, and Y. Wang. Cyclone: A Safe Dialect of C. In *Proceedings of the USENIX annual technical conference, Monterey, CA, June 2002.*, 2002. (Cited on pages 44.)

[129] Jim   Turley.   Embedded   Processors.   Available   at http://www.extremetech.com/article2/0,1558,18917,00.asp, 2002.   (Cited on pages 88.)

[130] S. P. Joglekar and S. R. Tate. Protomon: Embedded monitors for cryptographic protocol intrusion detection and prevention. In *Proceedings on the International Conference on Information Technology: Coding and Computing (ITCC'04)*, volume 1, page 81. IEEE Computer Society, 2004. (Cited on pages 53.)

[131] R. W. M. Jones and P. H. J. Kelly. Backwards-compatible bounds checking for arrays and pointers in c programs. In *Proceedings of the Automated and algorithmic debugging*, pages 13–26, 1997. (Cited on pages 64.)

[132] M. M. Kaempf. Vudo - an object superstitiously believed to embody magical powers. *Phrack Magazine*, 11(57), 2001. (Cited on pages 19.)

[133] Z. Kalbarczyk, R. K. Iyer, G. L. Ries, J. U. Patel, M. S. Lee, and Y. Xiao. Hierarchical simulation approach to accurate fault modeling for system dependability evaluation. *IEEE Transaction on Software Engineering*, 25(5):619–632, 1999. (Cited on pages 84.)

[134] G. Kanawati, V. Nair, N. Krishnamurthy, and J. Abraham. Evaluation of integrated system-level checks for on-line error detection. In *Proceedings of IEEE International Computer Performance and Dependability Symposium*, pages 292–301. IEEE Computer Society, September 1996. (Cited on pages 78, 147.)

[135] G. A. Kanawati, N. A. Kanawati, and J. A. Abraham. Ferrari: a tool for the validation of system dependability properties. In *Proceedings of The Twenty-Second Annual International Symposium on Fault-Tolerant Computing (FTCS-22)*, pages 336–344, 1992. (Cited on pages 84.)

[136] N. Kanawati, G. Kanawati, and J. Abraham. Dependability evaluation using hybrid fault/error injection. In *IPDS '95: Proceedings of the International Computer Performance and Dependability Symposium on Computer Performance and Dependability Symposium*, page 224, Washington, DC, USA, 1995. IEEE Computer Society. (Cited on pages 85.)

[137] J. Karlsson, P. Folkesson, J. Arlat, Y. Crouzet, G. Leber, and J. Reisinger. Application of Three Physical Fault Injection Techniques to the Experimental Assessment of the MARS Architecture. In *Proceedings of the Fifth IFIP Working Conference on Dependable Computing for Critical Applications (DCCA-5)*, pages 267–287, 1995. (Cited on pages 83.)

[138] J. Karlsson, U. Gunneflo, and J. Torin. The effects of heavy-ion induced single event upsets in the mc6809e microprocessor. In *Proceedings of Fehlertolerierende rechensysteme / fault-tolerant computing systems, automatisierungssysteme, methoden, anwendungen / automation systems, methods, applications; 4. internationale gi/itg/gma-fachtagung*, pages 296–307, London, UK, 1989. Springer-Verlag. (Cited on pages 33, 83.)

[139] J. Karlsson, P. Liden, P. Dahlgren, R. Johansson, and U. Gunneflo. Using heavy-ion radiation to validate fault-handling mechanisms. *IEEE Micro*, 14(1):8–11, 13–23, 1994. (Cited on pages 83.)

[140] G. S. Kc, A. D. Keromytis, and V. Prevelakis. Countering code-injection attacks with instruction-set randomization. In *Proceedings of the 10th acm conference on computer and communications security (Ccs '03)*, pages 272–280. ACM Press, 2003. (Cited on pages 43.)

[141] J. Kelsey, B. Schneier, D. Wagner, and C. Hall. Side channel cryptanalysis of product ciphers. In *Proceedings of the 5th European Symposium on Research in Computer Security*, pages 97–110, September 1998. (Cited on pages 100.)

[142] V. Kiriansky, D. Bruening, and S. P. Amarasinghe. Secure execution via program shepherding. In *Proceedings of the 11th USENIX Security Symposium*, pages 191–206, Berkeley, CA, USA, 2002. USENIX Association. (Cited on pages 13, 59, 60.)

[143] Klog. The Frame Pointer Overwrite. *Phrack Magazine*, 9(55), 1999. (Cited on pages 17, 62.)

[144] O. Kmmerling and M. G. Kuhn. Design principles for tamper-resistant smartcard processors. In *Proceedings of the USENIX workshop on smartcard technology (smart'99)*, pages 9–20, May 1999. (Cited on pages 100.)

[145] P. Kocher, J. Jaffe, and B. Jun. Differential power analysis. *Lecture Notes in Computer Science*, 1666:388–397, 1999. (Cited on pages 12.)

[146] P. Kocher, R. Lee, G. McGraw, A. Ragunanthan, and S. Ravi. Security as a new dimension in embedded system design, June 2004. (Cited on pages 5, 15.)

[147] S. Kowshik, D. Dhurjati, and V. Adve. Ensuring code safety without runtime checks for real-time control systems. In *Proceedings of the 2002 international conference on compilers, architecture, and synthesis for embedded systems (Cases '02)*, pages 288–297, New York, NY, USA, 2002. ACM Press. (Cited on pages 45.)

[148] A. Krennmair. ContraPolice: a libc extension for protecting applications from heap-smashing attacks. Available at http://www.synflood.at/contrapolice/, November 2003. (Cited on pages 68.)

[149] B. W. Lampson. Protection. In *Proceedings of the 5th Princeton Conferece on Information Sciences and Systems*, pages 18–24, Princeton, USA, 1971. ACM Operating Systems. (Cited on pages 55.)

[150] J. C. Laprie. Dependable Computing and Fault Tolerance: Basic Concepts and Terminology. In *Proceedings of the 15th International IEEE Symposium on Fault Tolerant Computing (FTCS-15)*, pages 2–11, Ann Arbor, MI, 1985. (Cited on pages 83.)

[151] D. Larochelle and D. Evans. Statically detecting likely buffer overflow vulnerabilities. In *Proceedings of the 10th USENIX Security Symposium*, pages 177–190, August 2001. (Cited on pages 45, 46.)

[152] E. Larson and T. Austin. High coverage detection of input-related security faults. In *Proceedings of the 12th Usenix security symposium*, Aug 2003. (Cited on pages 50.)

[153] J. R. Larus, T. Ball, M. Das, R. DeLine, M. Fähndrich, J. D. Pincus, S. K. Rajamani, and R. Venkatapathy. Righting software. *IEEE Software (magazine)*, 21(3):92–100, 2004. (Cited on pages 45, 46, 47.)

[154] D. Lea. A Memory Allocator. glibc-2.2.3/malloc/malloc.c. Comments in source code and available at http://g.oswego.edu/dl/html/malloc.html, 2000. (Cited on pages 19.)

[155] R. Lee, D. Karig, J. McGregor, and Z. Shi. Enlisting hardware architecture to thwart malicious code injection. In *Proceedings of the International Conference on Security in Pervasive Computing*. Springer Verlag LNCS, March 2003. (Cited on pages 42, 111.)

[156] A. Lesea, S. Drimer, J. Fabula, C. Carmichael, and P. Alfke. The Rosetta Experiment: Atmospheric Soft Error Rate testing in differing technology FPGAs. *IEEE Transactions on Device and Materials Reliability*, 5(3):317–328, 2005. (Cited on pages 6.)

[157] R. Leveugle, T. Michel, and G. Saucier. Design of Microprocessors with Built-in On-line Test. In *Proceedings of the 20th International Symposium on Fault-Tolerant Computing, 1990. FTCS-20*, pages 450–456, June 1990. (Cited on pages 75.)

[158] K. S. Lhee and S. J. Chapin. Type-assisted dynamic buffer overflow detection. In *Proceedings of the 11th USENIX Security Symposium*, pages 81–88, Berkeley, CA, USA, 2002. USENIX Association. (Cited on pages 65, 66.)

[159] S. Lin and D. J. Costello. *Error Control Coding: Fundamentals and Applications*. Prentice Hall, Englewood Cliffs, NJ, 1983. (Cited on pages 82.)

[160] Litchfield David. Defeating the Stack Based Buffer Overflow Prevention Mechanism of Microsoft Windows 2003 Server. Technical report, Microsoft Cooporation, 2003. (Cited on pages 62.)

[161] B. Littlewood and L. Strigini. Software reliability and dependability: A Roadmap. In *Proceedings of the conference on the Future of Software Engineering (LCSE '00)*, pages 175–188, New York, NY, USA, 2000. ACM Press. (Cited on pages 6.)

[162] Lu David J. Watchdog processors and structural integrity checking. *IEEE Transactions on Computers*, 31(7):681–685, 1982. (Cited on pages 72, 75, 111.)

[163] M. R. Lyu, editor. *Software Fault Tolerance*. John Wiley and Sons Ltd, 1995. (Cited on pages 10.)

[164] H. Madeira, M. Z. Rela, F. Moreira, and J. G. Silva. RIFLE: A general purpose pin-level fault injector. In *Proceedings of European dependable computing conference*, pages 199–216, 1994. (Cited on pages 83.)

[165] H. Madeira and J. Silva. On-line signature learning and checking: experimental evaluation. In *Proceedings in CompEuro '91. 'Advanced Computer Technology, Reliable Systems and Applications'. 5th Annual European Computer Conference*, pages 642–643, July 1991. (Cited on pages 74.)

[166] H. Madeira and J. G. Silva. Experimental evaluation of the fail-silent behavior in computers without error masking. In *Proceedings of the twenty-fourth international symposium on Fault-tolerant computing, 1994 (ftcs-24)*, pages 350–359, 1994. (Cited on pages 33.)

[167] A. Mahmood, A. Ersoz, and E. J. McCluskey. Concurrent system-level error detection using a Watchdog Processor. In *Proceedings of the 15th International Symposium on Fault Tolerant Computing*, pages 145–152, 1985. (Cited on pages 72.)

[168] A. Mahmood and E. J. McCluskey. Concurrent error detection using watchdog processors - a survey. *IEEE Transactions on Computers*, 37(2):160–174, 1988. (Cited on pages 10, 72.)

[169] A. Mahmood, E. J. McCluskey, and D. M. Andrews. Writing Executable Assertions to Test Flight Software. In *Proceedings of 18th Annual Asilomar Circuits and Systems Conference*, pages 262–266, 1984. (Cited on pages 72.)

[170] I. Majzik. Concurrent Error Detection Using Watchdog Processors. Technical report, Department of Measurement and Instrument Engineering, Technical University of Budapest, 1996. (Cited on pages 75.)

[171] S. Mangard. A Simple Power-Analysis (SPA) Attack on Implementations of the AES Key Expansion. In P. J. Lee and C. H. Lim, editors, *Proceedings of the 5th International Conference on Information Security and Cryptology (ICISC 2002)*, volume 2587 of *Lecture Notes in Computer Science*, pages 343–358. Springer, 2003. (Cited on pages 12.)

[172] P. Marwedel and C. Gebotys. Secure and safety-critical vs. insecure, non safety-critical embedded systems: do they require completely different design approaches? In *Proceedings of the 2nd ieee/acm/ifip international conference on hardware/software codesign and system synthesis (Codes+isss '04)*, pages 72–73. ACM Press, 2004. (Cited on pages 15.)

[173] Matt Conover and W00W00 Security Team. w00w00 on heap overflows. w00w00 Security Development (WSD), 1999. (Cited on pages 19.)

[174] Matt Messier and John Viega. Safe C String Library. Available at http://www.zork.org/safestr/, 2005. (Cited on pages 67.)

[175] T. May and M. Woods. Alpha particle induced soft errors in dynamic memories. *IEEE Transactions on Electronic Devices*, 26(2), 1979. (Cited on pages 6.)

[176] McFearin L. and Nair V.S.S. Control-flow checking using assertions. In *Proceedings of the 5th ifip international working conference on dependable computing for critical applications (dcca-5)*, pages 103–112. IEEE Computer Society Press, September 1995. (Cited on pages 78.)

[177] J. McGregor, D. Karig, Z. Shi, and R. Lee. A processor architecture defense against buffer overflow attacks. In *Proceedings of the International Conference on Security in Pervasive Computing (SPC-2003)*, pages 237–252. Springer Verlag, March 2003. (Cited on pages 41, 42.)

[178] A. Messer, P. Bernadat, D. D. Mannaru, G. Fu, D. Chen, Z. Dimitrijevic, D. Lie, A. Riska, and D. Milojicic. Susceptibility of commodity systems and software to memory soft errors. *IEEE Transactions on Computers*, 53(12):1557–1568, 2004. (Cited on pages 33.)

[179] E. Michel and W. Hohl. Concurrent error detection using watchdog processors in the multiprocessor system MEMSY. *Fault Tolerant Computing Systems, Springer Verlag*, 283:54–64, 1991. (Cited on pages 75.)

[180] T. Michel, R. Leveugle, F. Gaume, and R. Roane. An application specific microprocessor with two-level built-in control flow checking capabilities. In *Proceedings of the Euro ASIC'92*, pages 310–313. IEEE Computer Society, June 1992. (Cited on pages 75.)

[181] T. Michel, R. Leveugle, and G. Saucier. A new approach to control flow checking without program modification. In *Proceedigns of the twenty-first international symposium on fault-tolerant computing (FTCS-21)*, pages 334–341, June 1991. (Cited on pages 75.)

[182] Microsoft. Vault: A programming language for reliable systems. Available at http://research.microsoft.com/vault/, 2001. (Cited on pages 45.)

[183] M. Milenkovic. *Architectures for run-time verification of code integrity*. PhD thesis, The University of Alabama in Huntsville, Huntsville, AL, USA, 2005. Chair-Emil Jovanov. (Cited on pages 118, 136.)

[184] M. Milenkovic, A. Milenkovic, and E. Jovanov. A framework for trusted instruction execution via basic block signature verification. In *Proceedings of the 42nd annual southeast regional conference (ACM-SE 42)*, pages 191–196, New York, NY, USA, 2004. ACM Press. (Cited on pages 5, 126.)

[185] M. Milenkovic, A. Milenkovic, and E. Jovanov. Hardware support for code integrity in embedded processors. In *Proceedings of the 2005 international conference on compilers, architectures and synthesis for embedded systems (CASES '05)*, pages 55–65, New York, NY, USA, 2005. ACM Press. (Cited on pages 5, 15.)

[186] T. C. Miller. strlcpy and strlcat - consistent, safe, string copy and concatenation. In *Proceedings of 1999 USENIX Annual Technical Conference*, pages 175–178, Monterey, California, USA, 1999. USENIX Association. (Cited on pages 67.)

[187] D. Milojicic, A. Messer, J. Shau, G. Fu, and A. Munoz. Increasing relevance of memory hardware errors: a case for recoverable programming models. In *Proceedings of the 9th workshop on acm sigops european workshop (Ew 9)*, pages 97–102, New York, NY, USA, 2000. ACM Press. (Cited on pages 82.)

[188] G. Miremadi, J. Harlsson, U. Gunneflo, and J. Torin. Two software techniques for on-line error detection. In *Proceedings of the Twenty-second international symposium on fault-tolerant computing, 1992. ftcs-22*, pages 328–335. IEEE Computer Society, July 1992. (Cited on pages 78, 79.)

[189] S. Mitra, N. Seifert, M. Zhang, Q. Shi, and K. S. Kim. Robust system design with built-in soft-error resilience. *IEEE Computer (magazine)*, 38(2):43–52, 2005. (Cited on pages 6, 7, 32.)

[190] R. Muresan and C. H. Gebotys. Current flattening in software and hardware for security applications. In *Proceedings of the 2nd ieee/acm/ifip international conference on hardware/software codesign and system synthesis (Codes+isss '04)*, pages 218–223. ACM Press, 2004. (Cited on pages 12.)

[191] Nair V. S. S., Kim H., Krishnamurthy N., and Abraham J.A. Design and evaluation of automated high-level checks for signal processing applications. In *Proceedings of SPIE Advanced Algorithms and Architectures for Signal Processing Conference*, pages 292–301, August 1996. (Cited on pages 78.)

[192] N. Nakka, Z. Kalbarczyk, R. K. Iyer, and J. Xu. An architectural framework for providing reliability and security support. In *Proceedings of the 2004 International Conference on Dependable Systems and Networks*. IEEE Computer Society, 2004. (Cited on pages 39, 40, 77, 92.)

[193] N. Nakka, G. P. Saggese, Z. Kalbarczyk, and R. K. Iyer. An architectural framework for detecting process hangs/crashes. In *Proceedings of the 5th European Dependable Computing Conference, Budapest, Hungary, April 20-22, 2005*, pages 103–121, 2005. (Cited on pages 39, 40.)

[194] M. Namjoo. Techniques for concurrent testing of VLSI processor operation. In *Proceedings of the International Test Conference*, pages 461–468, 1982. (Cited on pages 74.)

[195] M. Namjoo and E. McCluskey. WATCHDOG PROCESSORS AND CAPABIL-ITY CHECKING. In *Proceedings of the Twenty-Fifth International Symposium on Fault-Tolerant Computing, 1995, ' Highlights from Twenty-Five Years'*, pages 94–, 1995. (Cited on pages 72.)

[196] Nathan P. Smith. Stack Smashing Vulnerabilities In The Unix Operating System. Available at http://www.cgsecurity.org/exploit/buffer-alt.ps, 1997. (Cited on pages 17.)

[197] G. C. Necula. Proof-Carrying Code. In *Proceedings of the 24th ACM SIGPLAN-SIGACT symposium on principles of programming languages (POPL '97)*, pages 106–119, Paris, France, Jan 1997. (Cited on pages 13.)

[198] G. C. Necula, J. Condit, M. Harren, S. McPeak, and W. Weimer. CCured: Type-safe Retrofitting of Legacy Software. *ACM Transactions on Programming Languages and Systems (TOPLAS)*, 27(3):477–526, 2005. (Cited on pages 44.)

[199] G. C. Necula, S. McPeak, S. P. Rahul, and W. Weimer. CIL: Intermediate Language and Tools for Analysis and Transformation of C Programs. In R. N. Horspool, editor, *Proceedings of the 11th International Conference on Compiler Construction, CC 2002, held as part of the joint European Conferences on Theory and Practice of Software, ETAPS 2002*, volume 2304 of *Lecture Notes in Computer Science*, pages 213–228, Grenoble, France, April 8-12, 2002, 2002. Springer. (Cited on pages 44.)

[200] G. C. Necula, S. McPeak, and W. Weimer. Ccured: type-safe retrofitting of legacy code. In *Proceedings of the 29th acm sigplan-sigact symposium on principles of programming languages*, pages 128–139, New York, NY, USA, 2002. ACM Press. (Cited on pages 44.)

[201] J. M. Nick, B. B. Moore, J.-Y. Chung, and N. S. Bowen. S/390 cluster technology: Parallel sysplex. *IBM Systems Journal*, 36(2):172–201, 1997. (Cited on pages 81.)

[202] G. Noubir and B. Choueiry. Algebraic techniques for the optimization of control flow checking. In *Proceedings of the annual symposium on fault tolerant computing*, pages 128–137. IEEE Computer Society, June 1996. (Cited on pages 10.)

[203] N. Oh, P. Shirvani, and E. McCluskey. Control-flow checking by software signatures. *IEEE Transactions on Reliability*, 51(1):111–122, 2002. (Cited on pages 80.)

[204] N. Oh, P. P. Shirvani, and E. J. McCluskey. Error Detection by Duplicated Instructions in Super-Scalar Processors. *IEEE Transactions on Reliability*, 51(1):63–75, 2002. (Cited on pages 80.)

[205] J. Ohlsson and M. Rimen. Implicit signature checking. In *Proceedins of the Twenty-fifth international symposium on fault-tolerant computing, 1995 (ftcs-25)*, pages 218–227. IEEE Computer Society Press, June 1995. (Cited on pages 74, 78.)

[206] J. Ohlsson, M. Rimen, and U. Gunneflo. A study of the effects of transient fault injection into a 32-bit RISC with built-in watchdog. In *Proceedings of the twenty-second international symposium on fault-tolerant computing, 1992. ftcs-22*, pages 316–325. IEEE Computer Society, July 1992. (Cited on pages 10, 33.)

[207] Y. Oiwa, T. Sekiguchi, E. Sumii, and A. Yonezawa. Fail-Safe ANSI-C compiler: An Approach to Making C Programs Secure. In *Proceedings of International Symposium on Software Security 2002*, 2002. (Cited on pages 65, 66.)

[208] S. M. Ornstein, W. R. Crowther, M. F. Kraley, and R. D. Bressler. Pluribus - A reliable multiprocessor. In *Proceedings of the AFIPS Conference*, volume 44, pages 551–559, 1975. (Cited on pages 72.)

[209] H. Ozdoganoglu, T. Vijaykumar, E. Brodley, A. Jalote, and A. Kuperman. Smash-Guard: A Hardware Solution to Prevent Security Attacks on the Function Return Address. Technical Report TR-ECE-03-13, Purdue University, 2002. (Cited on pages 42.)

[210] PaX Team. Documentation for the PaX project. Available at http://pax.grsecurity.net/docs/, 2005. (Cited on pages 71.)

[211] J. Peddersen, S. L. Shee, A. Janapsatya, and S. Parameswaran. Rapid Embedded Hardware/Software System Generation. In *Proceedings of the 18th International Conference on VLSI Design held jointly with 4th International Conference on Embedded Systems Design (VLSID'05)*, pages 111–116, January 2005. (Cited on pages 122, 138, 152.)

[212] Phishing Definition. Wikipedia Foundation Inc., the free encyclopedia,. http://en.wikipedia.org/wiki/Phishing, 2006. (Cited on pages 13.)

[213] J. Pincus and B. Baker. Beyond stack smashing: Recent advances in exploiting buffer overruns. *IEEE Security and Privacy (magazine)*, 2(4):20–27, 2004. (Cited on pages 30.)

[214] Pirate Decryption Definition. Wikipedia Foundation Inc., the free encyclopedia,. http://en.wikipedia.org/wiki/Pirate_decryption, 2006. (Cited on pages 12.)

[215] Poul-Henning Kamp. BSD HEAP SMASHING, May 2003. (Cited on pages 19.)

[216] D. Powell. Distributed fault tolerance: Lessons from delta-4. *IEEE Micro*, 14(1):36–47, 1994. (Cited on pages 83.)

[217] V. Prevelakis and D. Spinellis. Sandboxing applications. In *Proceedings of the FREENIX Track: 2001 USENIX Annual Technical Conference*, pages 119–126, Berkeley, CA, USA, 2001. USENIX Association. (Cited on pages 58.)

[218] N. Provos. Improving host security with system call policies. In *Proceedings of the 12th USENIX Security Symposium*, pages 257–272. USENIX Association, 2003. (Cited on pages 58, 59.)

[219] N. Quach. High Availability and Reliability in the Itanium Processor. *IEEE Micro*, 20(5):61–69, 2000. (Cited on pages 82.)

[220] J. J. Quisquater and D. Samyde. Side channel cryptanalysis. In *Proceedings of the SECI 2002*, pages 179–184, 2002. (Cited on pages 100.)

[221] R. G. Ragel and S. Parameswaran. Soft error detection and recovery in application specific instruction-set processors. In *Proceedings of the Workshop on System Effects of Logic Soft Errors*, April 2005. (Cited on pages 110.)

[222] R. G. Ragel, S. Parameswaran, and S. M. Kia. Micro embedded monitoring for security in application specific instruction-set processors. In *proceedings of the 2005 international conference on compilers, architectures and synthesis for embedded systems (CASES '05)*, pages 304–314, New York, NY, USA, 2005. ACM Press. (Cited on pages 110.)

[223] B. Ramamurthy and S. Upadhyaya. Watchdog processor-assisted fast recovery in distributed systems. In *Proceedings of the Fifth ieee international working conference on dependable computing for critical applications*, pages 125–134. IEEE Computer Society Press, September 1995. (Cited on pages 75.)

[224] J. R. Rao and P. Rohatgi. Empowering side-channel attacks. Cryptology ePrint Archive, Report 2001/037, 2001. (Cited on pages 12.)

[225] T. Rao. *Error Coding for Arithmetic Processors*. Academic Press, 1974. (Cited on pages 72.)

[226] F. Rashid, K. K. Saluja, and P. Ramanathan. Fault tolerance through re-execution in multiscalar architecture. In *Proceedings of the 2000 International Conference on Dependable Systems and Networks (formerly FTCS-30 and DCCA-8)*, pages 482–491. IEEE Computer Society Press, 2000. (Cited on pages 82.)

[227] S. Ravi, A. Raghunathan, and S. Chakradhar. Tamper resistance mechanisms for secure, embedded systems. In *Proceedings of the 17th international conference on vlsi design*, January 2004. (Cited on pages 12, 38, 39, 100.)

[228] S. Ravi, A. Raghunathan, P. Kocher, and S. Hattangady. Security in embedded systems: Design challenges. *ACM Transactions on Embedded Computing Systems*, 3(3):461–491, 2004. (Cited on pages 2, 12, 13, 15.)

[229] J. Ray, J. C. Hoe, and B. Falsafi. Dual use of superscalar datapath for transient-fault detection and recovery. In *Proceedings of the 34th annual ACM/IEEE international symposium on Microarchitecture*, pages 214–224. IEEE Computer Society, 2001. (Cited on pages 82.)

[230] M. Rebaudengo, M. S. Reorda, M. Torchiano, and M. Violante. Soft-error detection through software fault-tolerance techniques. In *Proceedings in the 14th International Symposium on Defect and Fault-Tolerance in VLSI Systems (DFT '99)*, 0-7695-0325-X, pages 210–218, Albuquerque, NM, USA, 1999. (Cited on pages 82.)

[231] S. K. Reinhardt and S. S. Mukherjee. Transient fault detection via simultaneous multithreading. In *Proceedings of the 27th annual international symposium on Computer architecture*, pages 25–36. ACM Press, 2000. (Cited on pages 6, 82.)

[232] G. A. Reis, J. Chang, N. Vachharajani, R. Rangan, and D. I. August. SWIFT: Software Implemented Fault Tolerance. In *Proceedings of the international symposium on code generation and optimization (CGO '05)*, pages 243–254, Washington, DC, USA, 2005. IEEE Computer Society. (Cited on pages 80.)

[233] G. A. Reis, J. Chang, N. Vachharajani, R. Rangan, D. I. August, and S. S. Mukherjee. Software-controlled fault tolerance. *ACM Transactions on Architecture and Code Optimization (TACO)*, 2(4):366–396, 2005. (Cited on pages 80.)

[234] M. Z. Rela, H. Madeira, and J. G. Silva. Experimental evaluation of the fail-silent behaviour in programs with consistency checks. In *Proceedings of the the twenty-sixth annual international symposium on fault-tolerant computing (ftcs '96)*, page 394, Washington, DC, USA, 1996. IEEE Computer Society. (Cited on pages 82.)

[235] G. Richarte. Four different tricks to bypass stackshield and stackguard protection, April 2002. (Cited on pages 60, 61, 62.)

[236] G. L. Ries. *Hierarchical simulation to assess hardware and software dependability*. PhD thesis, University of Illinois at Urbana-Champaign, 1997. Adviser-Ravishankar K. Iyer. (Cited on pages 84.)

[237] RISC Definition. Wikipedia Foundation Inc., the free encyclopedia,. Available at http://en.wikipedia.org/wiki/RISC, 2006. (Cited on pages 88.)

[238] rix. Smashing c++ vptrs. *Phrack Magazine*, 10(56), 2000. (Cited on pages 19.)

[239] Robert H. Morelos-Zaragoza. *The Art of Error Correcting Coding*. John Wiley and Sons, first edition edition, 2002. (Cited on pages 82.)

[240] W. Robertson, C. Kruegel, D. Mutz, and F. Valeur. Run-time Detection of Heap-based Overflows. In *Proceedings of the 17th USENIX Conference on System Administration (LISA '03)*, pages 51–60, Berkeley, CA, USA, 2003. USENIX Association. (Cited on pages 68, 69.)

[241] E. Rotenberg. AR-SMT: A microarchitectural approach to fault tolerance in microprocessors. In *Proceedings of the Symposium on fault-tolerant computing*, pages 84–91, 1999. (Cited on pages 82.)

[242] P. Rubinfeld. Virtual roundtable on the challenges and trends in processor design: Managing problems at high speeds. *IEEE Transactions on Computers*, January 1998. (Cited on pages 6.)

[243] R. Rugina and M. C. Rinard. Symbolic bounds analysis of pointers, array indices, and accessed memory regions. In *Proceedings of the SIGPLAN Conference on Programming Language Design and Implementation*, pages 182–195, 2000. (Cited on pages 48.)

[244] R. Rugina and M. C. Rinard. Symbolic bounds analysis of pointers, array indices, and accessed memory regions. *ACM Transactions on Programming Languages and Systems (TOPLAS)*, 27(2):185–235, 2005. (Cited on pages 48, 49.)

[245] O. Ruwase and M. Lam. A practical dynamic buffer overflow detector. In *Proceedings of the Network and Distributed System Security (NDSS) Symposium*, pages 159–169, 2004. (Cited on pages 64.)

[246] R. A. Sahner, K. S. Trivedi, and A. Puliafito. *Performance and reliability analysis of computer systems: an example-based approach using the SHARPE software package*. Kluwer Academic Publishers, Norwell, MA, USA, 1996. (Cited on pages 83.)

[247] J. H. SALTZER and M. D. SCHROEDER. The Protection of Information in Computer Systems. In *Proceedings of the IEEE*, pages 1278–1308, September 1975. (Cited on pages 54.)

[248] J. R. Samson Jr., W. Moreno, and F. Falquez. A technique for automated validation of fault tolerant designs using laser fault injection (lfi). In *Proceedings of the the twenty-eighth annual international symposium on fault-tolerant computing (Ftcs '98)*, page 162, Washington, DC, USA, 1998. IEEE Computer Society. (Cited on pages 83.)

[249] N. Saxena and E. McCluskey. Control-flow checking using watchdog assists and extended-precision checksums. *IEEE Transactions on Computers*, pages 554–558, April 1990. (Cited on pages 74.)

[250] SCA Definition. Wikipedia Foundation Inc., the free encyclopedia,. http://en.wikipedia.org/wiki/Static_code_analysis, 2006. (Cited on pages 45.)

[251] M. Schmid, R. Trapp, A. Davidoff, and G. Masson. Upset Exposure by Means of Abstraction Verification. In *Proceedings of the 12th IEEE Fault-Tolerant Computing Symposium*, pages 237–244, 1982. (Cited on pages 33.)

[252] F. Schneider. Least privilege and more. *IEEE Security and Privacy (magazine)*, 1(5):55–59, 2003. (Cited on pages 54.)

[253] F. B. Schneider. Enforceable security policies. *ACM Transactions on Information and System Security (TISSEC)*, 3(1):30–50, 2000. (Cited on pages 56.)

[254] M. A. Schuette and J. P. Shen. Processor control flow monitoring using signatured instruction streams. *IEEE Transactions on Computers*, 36(3):264–276, 1987. (Cited on pages 10, 75.)

[255] M. A. Schuette, J. P. Shen, D. P. Siewiorek, and Y. X. Zhu. Experimental evaluation of two concurrent error detection schemes. In *Digestions of papers of the 16th annual international symposium of fault-tolerant computing (FTCS-16)*, pages 138–143, July 1986. (Cited on pages 75.)

[256] scut and Team Teso. Exploiting Format String Vulnerabilities. Available at http://koti.welho.com/vskytta/formatstring-1.2.pdf, September 2001. (Cited on pages 29.)

[257] Secure Software. Secure Software RATS System. Available at http://www.securesoftware.com/resources/, 2004. (Cited on pages 50.)

[258] Z. Segall, D. Vrsalovic, D. Siewiorek, D. Yaskin, J. Kownacki, J. Barton, R. Dancey, A. Robinson, and T. Lin. FIAT: Fault-injection based automated testing environment. In *Proceedings of the 18th International Symposium on Fault-Tolerant Computing*, pages 102–107, 1988. (Cited on pages 84.)

[259] N. Seifert, P. Slankard, M. Kirsch, B. Narasimham, V. Zia, C. Brookreson, A. Vo, S. Mitra, and J. Maiz. Radiation Induced Soft Error Rates of Advanced CMOS Bulk Devices. In *Proceedings of the IEEE International Reliability Physics Symposium*, 2006. (Cited on pages 6, 8.)

[260] R. Sekar, M. Bendre, D. Dhurjati, and P. Bollineni. A fast automaton-based method for detecting anomalous program behaviors. In *Proceedings of the 2001 ieee symposium on security and privacy*, page 144, Washington, DC, USA, 2001. IEEE Computer Society. (Cited on pages 53.)

[261] SEU Definition. Wikipedia Foundation Inc., the free encyclopedia,. http://en.wikipedia.org/wiki/Single_event_upset, 2006. (Cited on pages 6, 32.)

[262] U. Shankar, K. Talwar, J. Foster, and D. Wagner. Detecting format string vulnerabilities with type qualifiers. In *Proceedings of the 10th USENIX Security Symposium, 2001.*, 2001. (Cited on pages 46, 47.)

[263] Z. Shao, C. Xue, Q. Zhuge, E. H.-M. Sha, and B. Xiao. Security protection and checking in embedded system integration against buffer overflow attacks. In *Proceedings of International conference on information technology: coding and computing, las vegas, nevada*. IEEE Computer Society, April 2004. (Cited on pages 42.)

[264] Z. Shao, Q. Zhuge, Y. He, and E. H. M. Sha. Defending embedded systems against buffer overflow via hardware/software. In *Proceedings of the 19th Annual Computer Security Applications Conference*, page 352. IEEE Computer Society, 2003. (Cited on pages 42.)

[265] P. Shivakumar, S. W. Keckler, C. R. Moore, and D. Burger. Exploiting microarchitectural redundancy for defect tolerance. In *Proceedings of the 21st international conference on computer design (Iccd '03)*, page 481, Washington, DC, USA, 2003. IEEE Computer Society. (Cited on pages 8.)

[266] P. Shivakumar, M. Kistler, S. Keckler, D. Burger, and L. Alvisi. Modeling the effect of technology trends on the soft error rate of combinational logic. In *Proceedings in the International Conference on Dependable Systems and Networks (DSN'02)*, pages 389 – 398, 2002. (Cited on pages 32.)

[267] D. Siewiorek and L. K.-W. Lai. Testing of digital systems. In *Proceedings of the IEEE*, pages 1321–1333, 1981. (Cited on pages 10, 32.)

[268] D. Siewiorek and R. Swarz. *Theory and Practice of Reliable System Design*. Digital Press, 1982. (Cited on pages 10, 32.)

[269] A. Simon and A. King. Analyzing String Buffers in C. In H. Kirchner and C. Ringeissen, editors, *Proceedings of the International conference on algebraic methodology and software technology*, volume 2422 of *Lecture Notes in Computer Science*, pages 365–379. Springer, September 2002. (Cited on pages 46.)

[270] K. Skadron, P. S. Ahuja, M. Martonosi, and D. W. Clark. Improving prediction for procedure returns with return-address-stack repair mechanisms. In *Proceedings of the 31st annual acm/ieee international symposium on microarchitecture (Micro 31)*, pages 259–271, Los Alamitos, CA, USA, 1998. IEEE Computer Society Press. (Cited on pages 41, 42.)

[271] C. Small. A tool for constructing safe extensible C++ systems. In *Proceedings of the 4th USENIX Conference on Object-Oriented Technologies and Systems (COOTS)*, pages 175–184, 1997. (Cited on pages 55, 56.)

[272] A. Snarskii. FreeBSD libc stack integrity patch. Available at ftp://ftp.lucky.net/pub/unix/local/libc-letter, February 1997. (Cited on pages 68.)

[273] Soft Error Definition. Wikipedia Foundation Inc., the free encyclopedia,. http://en.wikipedia.org/wiki/Soft_error, 2006. (Cited on pages 32.)

[274] J. Sosnowski. Detection of control flow errors using signature and checking instructions. *Proceedings of the IEEE International Test Conference*, pages 81–88, 1988. (Cited on pages 74.)

[275] J. Sosnowski and J. Nowicki. Experiments with on-chip monitoring in pentium processors. Technical report, Institute of Computer Science, Warsaw University of Technology, 1998. (Cited on pages 77.)

[276] SRAM Definition. Wikipedia Foundation Inc., the free encyclopedia,. Available at http://en.wikipedia.org/wiki/Static_Random_Access_Memory, 2006. (Cited on pages 6.)

[277] T. Sridhar and S. M. Thatte. Concurrent Checking of Program Flow in VLSI Processors. In *Proceedings of the 12th IEEE International Test Conference*, pages 191–199, 1982. (Cited on pages 72, 74.)

[278] F.-X. Standaert, E. Peeters, and J.-J. Quisquater. On the masking countermeasure and higher-order power analysis attacks. In *Proceedings of the International Symposium on Information Technology: Coding and Computing (ITCC 2005)*, volume 1, page 406, Las Vegas, Nevada, USA, 2005. (Cited on pages 12.)

[279] J. L. Steffen. Adding run-time checking to the portable C compiler. *Software: Practice and Experience*, 22(4):305–316, 1992. (Cited on pages 64.)

[280] D. T. Stott, G. Ries, M.-C. Hsueh, and R. K. Iyer. Dependability analysis of a high-speed network using software-implemented fault injection and simulated fault injection. *IEEE Transactions on Computers*, 47(1):108–119, 1998. (Cited on pages 84, 85.)

[281] G. Suh, D. Clarke, B. Gassend, M. van Dijk, and S. Devadas. Hardware mechanisms for memory integrity checking. Technical report, MITLCS, November 2002. (Cited on pages 37, 38.)

[282] G. Suh, D. Clarke, B. Gassend, M. van Dijk, and S. Devadas. AEGIS: architecture for tamper-evident and tamper-resistant processing. In *Proceedings of the 17 international conference on supercomputing*, 2003. (Cited on pages 38.)

[283] SuperScalar Definition. Wikipedia Foundation Inc., the free encyclopedia,. Available at http://en.wikipedia.org/wiki/Superscalar, 2006. (Cited on pages 159.)

[284] B. D. Sutter, B. D. Bus, and K. D. Bosschere. Link-time binary rewriting techniques for program compaction. *ACM Transactions on Programming Languages and Systems (TOPLAS)*, 27(5):882–945, 2005. (Cited on pages 118, 136.)

[285] Technical Brief. Hot Plug RAID Memory Technology for Fault Tolerance and Scalability. Technical report, Hewlett Packard, 2002. (Cited on pages 82.)

[286] The Altera Team. Using Parity to Detect Errors. Technical report, Altera, The Programmable Solutions Company, 2005. (Cited on pages 10.)

[287] The ARM Team. An introduction to thumb, advanced {RISC} machines ltd. Available at http://www.arm.com, March 1995. (Cited on pages 103.)

[288] The ARM Team. Arm Reference Manual, Advanced RISC Machines Ltd. Available at http://www.arm.com, July 2000. (Cited on pages 103.)

[289] The GCC Team. GNU/GCC Compiler, Free Software Foundation. Available at http://gcc.gnu.org. (Cited on pages 103.)

[290] The Mentor Graphics Team. Model Sim from Mentor Graphics, A hardware simulation tool. Available at http://www.model.com/. (Cited on pages 103.)

[291] The NetBSD Team. Non-executable stack and heap. Technical report, The NetBSD Foundation, Inc. Available at http://www.netbsd.org/Documentation/kernel/non-exec.html, 2005. (Cited on pages 71.)

[292] The Open Group. Rationale for International Standard Programming Language C. Available at http://std.dkuug.dk/JTC1/SC22/WG14/www/C99RationaleV5.10.pdf, April 2003. (Cited on pages 25.)

[293] The PEAS Team. ASIP Meister, Available at http://www.eda-meister.org/asip-meister/, 2002. (Cited on pages 88, 103, 122, 138, 153.)

[294] The SANS Team. The SANS Institute, The SANS/FBI Twenty Most Critical Internet Security Vulnerabilities, October 2004. (Cited on pages 95.)

[295] The Synopsys Team. Synopsys Design Compiler, The industry standard for logic synthesis. Available at http://www.synopsys.com/. (Cited on pages 103.)

[296] K. S. Trivedi. *Probability and Statistics with Reliability, Queuing and Computer Science Applications*. Prentice Hall PTR, Upper Saddle River, NJ, USA, 1982. (Cited on pages 83.)

[297] T. K. Tsai, R. K. Iyer, and D. Jewitt. An approach towards benchmarking of fault-tolerant commercial systems. In *Proceedings of the the twenty-sixth annual international symposium on fault-tolerant computing (ftcs '96)*, page 314, Washington, DC, USA, 1996. IEEE Computer Society. (Cited on pages 84.)

[298] S. Upadhyaya and B. Ramamurthy. Concurrent process monitoring with no reference signatures. *IEEE Transactions on Computers*, 43:475–480, April 1994. (Cited on pages 74.)

[299] Vendicator. Stack Shield: A "stack smashing" technique protection tool for Linux. Available at http://www.angelfire.com/sk/stackshield/, 2000. (Cited on pages 60.)

[300] A. Vetteth. Hardware Implementation of Reconfigurable Modules for Reliability and Security Engine. Master's thesis, Coordinated Science Laboratory, University of Illinois at Urbana-Champaign, May 2005. (Cited on pages 39, 107.)

[301] J. Viega, J. T. Bloch, Y. Kohno, and G. McGraw. ITS4: a static vulnerability scanner for c and c++ code. In *Proceedings of the 16th annual computer security applications conference (acsac 2000)*, pages 257–267, 2000. (Cited on pages 50.)

[302] T. N. Vijaykumar, I. Pomeranz, and K. Cheng. Transient-fault recovery using simultaneous multithreading. In *Proceedings of the 29th annual international symposium on Computer architecture*, pages 87–98. IEEE Computer Society, 2002. (Cited on pages 82.)

[303] VLIW Definition.   Wikipedia Foundation Inc., the free encyclopedia,. http://en.wikipedia.org/wiki/VLIW, 2006. (Cited on pages 160.)

[304] D. Wagner and D. Dean. Intrusion detection via static analysis. In *Proceedings of the 2001 ieee symposium on security and privacy (Sp '01)*, page 156, Washington, DC, USA, 2001. IEEE Computer Society. (Cited on pages 53.)

[305] D. Wagner, J. S. Foster, E. A. Brewer, and A. Aiken. A first step towards automated detection of buffer overrun vulnerabilities. In *Proceedings of the Network and Distributed System Security Symposium*, pages 3–17, San Diego, CA, February 2000. (Cited on pages 47, 48.)

[306] R. Wahbe, S. Lucco, T. E. Anderson, and S. L. Graham. Efficient software-based fault isolation. In *Proceedings of the fourteenth acm symposium on operating systems principles (Sosp '93)*, pages 203–216, New York, NY, USA, 1993. ACM Press. (Cited on pages 55.)

[307] C. Warrender, S. Forrest, and B. Pearlmutter. Detecting intrusions using system calls: alternative data models. In *Proceedings of the 1999 IEEE Symposium on Security and Privacy 1999*, pages 133–145, 1999. (Cited on pages 53.)

[308] C. Weaver and T. M. Austin. A fault tolerant approach to microprocessor design. In *Proceedings of the 2001 International Conference on Dependable Systems and Networks (formerly: FTCS)*, pages 411–420. IEEE Computer Society, 2001. (Cited on pages 82.)

[309] C. F. Webb. Subroutine call/return stack. *BM Technical Disclosure Bulletin*, 30(11), April 1988. (Cited on pages 41.)

[310] D. A. Wheeler. FlawFinder. Available at http://www.dwheeler.com/flawfinder/, 2001. (Cited on pages 50.)

[311] J. Wilander and M. Kamkar. A comparison of publicly available tools for static intrusion prevention. In *Proceedings of the 7th Nordic Workshop on Secure IT Systems*, pages 68–84, Karlstad, Sweden, November 2002. (Cited on pages 72.)

[312] J. Wilander and M. Kamkar. A comparison of publicly available tools for dynamic buffer overflow prevention. In *Proceedings of the 10th Network and Distributed*

System Security Symposium, pages 149–162, San Diego, California, February 2003. (Cited on pages 72.)

[313] K. Wilken and J. Shen. Continuous signature monitoring: low-cost concurrent detection of processor control errors. *IEEE Transactions on Computer-Aided Design of Integrated Circuits and Systems*, pages 629–641, June 1990. (Cited on pages 75.)

[314] K. D. Wilken. Optimal signature placement for process-error detection using signature monitoring. In *Proceedings of the 1991 International Symposium on Fault-Tolerant Computing*, pages 326–333, 1991. (Cited on pages 74.)

[315] K. D. Wilken. An optimal graph-construction approach to placing program signatures for signature monitoring. *IEEE Transactions on Computers*, 42(11):1372–1381, 1993. (Cited on pages 74.)

[316] K. D. Wilken and T. Kong. Concurrent detection of software and hardware data-access faults. *IEEE Transactions on Computers*, 46(4):412–424, 1997. (Cited on pages 76, 77.)

[317] William Stallings. *Cryptography and Network Security: Principles and Practice*. Prentice Hall, third edition edition, 1998. (Cited on pages 3.)

[318] R. Wojtczuk. Defeating Solar Designer's Non-executable Stack Patch. Available at http://www.insecure.org/sploits/non-executable.stack.problems.html, 1998. (Cited on pages 71.)

[319] G. Wurster, P. C. van Oorschot, and A. Somayaji. A generic attack on checksumming-based software tamper resistance. In *Proceedings of the 2005 ieee symposium on security and privacy*, pages 127–138, Washington, DC, USA, 2005. IEEE Computer Society. (Cited on pages 111.)

[320] Y. Xie, A. Chou, and D. Engler. Archer: using symbolic, path-sensitive analysis to detect memory access errors. In *Proceedings of the 9th european software engineering conference held jointly with 11th acm sigsoft international symposium on foundations of software engineering*, pages 327–336, New York, NY, USA, 2003. ACM Press. (Cited on pages 48.)

[321] J. Xu. Intrusion prevention using control data randomization. In *Proceedings of Supplementary of IEEE International conference on dependable systems and networks (DSN), San francisco, ca*, June 2003. (Cited on pages 71.)

[322] J. Xu, Z. Kalbarczyk, and R. K. Iyer. Transparent runtime randomization for security. In *Proceedings of the 22nd international symposium on reliable distributed systems, florence, italy*. IEEE Computer Society, October 2003. (Cited on pages 71.)

[323] J. Xu, Z. Kalbarczyk, S. Patel, and R. Iyer. Architecture support for defending against buffer overflow attacks. In *Proceedings of the Easy-2 workshop*, October 2002. (Cited on pages 41.)

[324] S. S. Yau and F.-C. Chen. An approach to concurrent control flow checking. *IEEE Transactions on Software Engineering*, 6(2):126–137, 1980. (Cited on pages 72, 75, 82.)

[325] S. H. Yong and S. Horwitz. Protecting c programs from attacks via invalid pointer dereferences. In *Proceedings of the 9th european software engineering conference held jointly with 11th acm sigsoft international symposium on foundations of software engineering (Esec/fse-11)*, pages 307–316, New York, NY, USA, 2003. ACM Press. (Cited on pages 63.)

[326] Y. Younan. An overview of common programming security vulnerabilities and possible solutions. Master's thesis, Vrije Universiteit Brussel, August 2003. (Cited on pages 25.)

[327] Y. Younan, W. Joosen, and F. Piessens. Code injection in C and CPP: A Survey of Vulnerabilities and Countermeasure. Technical report, Depertment of Computer Science, Katholieke Universiteit Leuven, Belgium, July 2004. (Cited on pages 45.)

[328] L. Young, C. Alonso, R. Iyer, and K. Goswami. A hybrid monitor assisted fault injection environment. In *Proceedings of Dependable Computing for Critical Applications*, pages 281–302, 1993. (Cited on pages 85.)

[329] L. Yount. Architectural solutions to safety problems of digital flight Critical systems for commercial transports. In *Proceedings of the Sixth Digital Avionics Systems Conference, Baltimore, MD*, pages 28–34, New York, 1984. American Institute of Aeronautics and Astronautics. (Cited on pages 73.)

[330] J. F. Ziegler and W. A. Lanford. The Effect of Cosmic Rays on Computer Memories. *Science*, 206:776, 1979. (Cited on pages 6.)